T0135626

Gas Distribution Mapping and Gas Source Localisation with a Mobile Robot

Dissertation

der Fakultät für Informations- und Kognitionswissenschaften
der Eberhard-Karls-Universität Tübingen
zur Erlangung des Grades eines
Doktors der Naturwissenschaften
(Dr. rer. nat.)

vorgelegt von

Dipl.-Phys. Achim Lilienthal

aus Meßkirch

Tübingen
2004

Bibliografische Information Der Deutschen Bibliothek

Die Deutsche Bibliothek verzeichnet diese Publikation in der Deutschen Nationalbibliografie; detaillierte bibliografische Daten sind im Internet über http://dnb.ddb.de abrufbar.

ISBN 3-8325-0790-6

Logos Verlag Berlin
Comeniushof, Gubener Str. 47,
10243 Berlin
Tel.: +49 030 42 85 10 90
Fax: +49 030 42 85 10 92
INTERNET: http://www.logos-verlag.de

Tag der mündlichen Qualifikation:	8. 12. 2004
Dekan:	Prof. Dr. Michael Diehl
1. Berichterstatter:	Prof. Dr. Andreas Zell
2. Berichterstatter:	Prof. Dr. Paolo Dario (Pisa)
3. Berichterstatter:	Prof. Dr.-Ing. Dr.-Ing. E.h. Wolfgang Straßer

Abstract

This thesis addresses two fundamental problems concerning the application of electrochemical gas sensors on a mobile robot in a real world environment: gas distribution mapping and localisation of a static gas source.

Apart from the limitations of current sensor technology, the main difficulty for both of these tasks results from the spreading of gases under natural conditions. Diffusion plays only a minor role in the distribution of odourant molecules at room temperature. The dominant transport mechanisms are convection flow and turbulence. As a consequence, the resulting concentration field is patchy and chaotic, and the actual gas source is usually not located at the point of highest concentration.

In order to create a map of a gas distribution it is thus required to extract time-constant properties of the concentration field from a series of sensor readings collected by a mobile robot. In contrast to conventional mapping techniques, where a considerable overlap between single measurements can be assumed, gas sensor measurements provide information about a very small area. The problem of gas distribution mapping is therefore to create a representation of the average concentration field from sparse point samples of the instantaneous distribution with little or no overlap between single measurements. A new algorithm is introduced in this work to create a representation that stores belief about the average relative concentration of a detected gas in a grid structure.

The problem of gas source localisation can be broken down into three subtasks: gas finding, source tracing and source declaration. Gas finding – the detection of an increased concentration – amounts to a basic search task and the selection of a suitable threshold value, a problem that is not addressed in this work. The main part of this thesis is rather concerned with the latter two aspects: following the cues determined from the sensed gas distribution towards its origin (source tracing) and establishing that the source has been found (source declaration). In order to avoid limitation to an environment with a strong airflow, the investigated strategies do not rely on information about wind direction and speed. Accordingly, the experiments were carried out in an unventilated indoor environment where previously suggested methods for gas source localisation, which include periods of upwind movement, are not applicable due to the limitations of current anemometers.

A straightforward solution for gas source tracing is to follow an instantaneously measured spatial concentration gradient. Such a strategy was realised on a mobile robot by implementing a direct sensor motor coupling in the manner of a Braitenberg vehicle. The particular contribution of this work is a detailed statistical evaluation of the tracing performance based on a large number of experiments. While gradient following is often mislead by transient concentration maxima, it could be shown with high statistical significance that the path length required to reach the source can be reduced on average compared to random search.

As a further gas source tracing method, a biomimetic strategy was investigated that is based on the key elements of the behaviour of male silkworm moths to find a mate guided by sexual pheromones. A modification for use on mobile robots is proposed that does not depend on information about the local wind speed. The performance of this tracing strategy was tested in a largely uncontrolled indoor environment and evaluated by statistical means. Despite the considerably larger size of the robot compared to the moth, the results suggest that the modified strategy decreases the average robot-to-source distance compared to random exploration, because it can keep the robot in the vicinity of a gas source after single gas patches have been discovered by initial exploration.

In order to address the full gas source localisation problem, a tracing strategy has to be extended by an additional declaration mechanism to determine that the source has been found with high certainty. It is generally not sufficient to search for maxima of the instantaneous concentration distribution in order to solve the declaration task. However, it was demonstrated that peaks in sensor response can provide a rough estimate of the gas source location if the search space is restricted. In experiments in a one dimensional scenario, the sensing strategy was found to have a dominant influence. A strong correlation between sensor response and proximity to a source could be obtained only if the robot was driven with a constant, sufficiently high speed, while such a correlation could not be observed if a stop-sense-go strategy was applied. A possible explanation for this effect suggests an alternative feature that might be used to recognise a gas source. Rather than looking for a global maximum of concentration, a gas source can be distinguished by an increased frequency of local maxima. This was also found in experiments in a two dimensional scenario where the sensor-motor connections of the implemented gas-sensitive Braitenberg vehicle were crossed. In this way, the robot performs exploration and concentration peak avoidance, resulting in a path that reflects the frequency of local maxima in the inspected area. A visualisation of this path offers therefore a possible method for gas source declaration without using additional sensors.

Finally, the thesis investigates the possibility to classify a suspected object as being a gas source or not from a pattern in a series of spatially and temporally sampled concentration data. Such a pattern was determined by applying machine learning techniques. The results of this ongoing work demonstrate the feasibility of the approach and show that high classification rates can be achieved using support vector machines. An analysis of the most important features for classification, the dependency of the classification rate on the desired declaration accuracy, and a comparison with the classification rate that can be achieved by selecting an optimal threshold value regarding the mean sensor signal is also presented.

Acknowledgements

First of all I am indebted to my supervisor, Prof. Dr. Andreas Zell, for providing a supportive research environment within his group that enabled me to concentrate on the real research work rather than keeping me busy with "unnecessary" things like obtaining funds or struggling with inadequate robots and computer equipment. He was brave enough to hire me although my proven skills in robotics were severely limited at that time, basically to the myriads of cardboard robots I used to build as a child. I would also like to thank my external examiners, Prof. Dr. Paolo Dario and Prof. Dr. Wolfgang Straßer for their friendly assessment of this work. A special thank you goes to Paolo Dario's secretary Lisa Puzella, who never grew tired of sympathetically explaining Paolo Dario's schedule to me and gave me exact instructions of how to catch him personally in Japan.

Basic research requires a certain amount of money that does not ask for a market-ready product as the first priority. I gratefully acknowledge such financial support from different institutions. My position at Tübingen University, where the main part of this thesis developed, was funded by the state of Baden-Württemberg. A further important contribution arose during a pleasant and highly productive six month stay at the Center of Applied Autonomous Sensor Systems at Örebro University in Sweden. This was made possible by a Marie Curie scholarship that was part of the European Commission's 5th Framework Programme. Ultimately, I would also like to extend this acknowledgement to the Deutsche Forschungsgemeinschaft that financially supported the trip to the IROS 2004 conference in Japan.

Next to my doctoral father Andreas Zell, Tom Duckett became a sort of "doctoral brother" to me. We were able to develop a virtually frictionless process of iterative paper writing, which could be demonstrated to work by personal communication as well as by telephone and Email. Thanks Tom for countless proofreadings, prolific discussions, and the good time we had while getting lost in various countries.

This thesis derives considerable benefit from good teamwork with my colleagues in Tübingen and Örebro. Most of the experiments with the gas sensitive robot "Arthur" (introduced in Chapter 3) were carried out together with the Institute of Physical Chemistry, namely with Udo Weimar and Michael Wandel to whom I want to say thank you for the friendly and stimulating cooperation. With Michael I shared many hours of "robot sitting" and he was willing to help with technical problems at any time. (I suppose not too many other people have had the chance to enjoy a

v

detailed analysis of some inaccurate radiator welding at 5 o'clock in the morning.)
Moreover, this work would possibly deal with a different topic, if he hadn't suggested
to speed up the robot "just for fun" at the very moment we had almost decided to
stop the gas source localisation experiments in a corridor-like environment because
the approach (which is detailed in Chapter 4) seemed to have failed completely.
Instead, the decision to record concentration measurements while driving became
the actual starting point of the results presented here by providing the missing drop
of knowledge, which was necessary to go beyond the required critical mass.

Furthermore, I would like to thank the students Felix Werner, Denis Reiman, and
Andreas Stützle, for their open-minded passion and the many hours of spare time
they dedicated to the gas-sensitive mobile robots whilst writing their B. Sc. and
M. Sc. theses, respectively.

Aside from direct cooperation in the field of mobile noses, there are several fellow
PhD students I want to say thank you to. To Holger Ulmer my congenial car-
pool partner. To my roomates Michael Plagge and Patrick Heinemann, with whom
I enjoyed a very pleasant working atmosphere. To Amy Loutfi, Holger Fröhlich,
Christian Spieth, Felix Streichert, André Treptow (and again Patrick Heinemann,
Holger Ulmer, and Michael Wandel) for reading and commenting on my papers and
previous drafts of this thesis. To Boyko Iliev, Kevin LeBlanc, the "Koala chief en-
gineer" Alexander Skoglund, and Per Sporrong, for their kind help when setting up
the experiments in Örebro. To Dietmar Krieg for his help on aerodynamics, and
Jan Pohland and Ivan Kalaykov for their advice on statistics questions. To Grzegorz Cielniak who helped to determine the exact centre of the universe and who
was an excellent hat model no matter whether he was wearing a bright green or a
stylish blue hat. Finally, to my loyal companion LEO[1] who was always willing to
give support concerning the English language.

Ultimately, I want to express my gratitude to the PhD experience itself for providing me with an opportunity to get to know great people I probably would never
have met otherwise. Without underestimating all the others, I would like to particularly mention Simon Wiest, Tom Duckett, Grzegorz Cielniak and Amy Loutfi with
whom I developed a special friendship over the past years. And finally I want to
thank my dear friend Dr. Wolfi for his scientific advice, my brother Ralph for his
valuable pedagogical comments, my family for their unquestioning support, and the
great person who came along with me through all the years: Tanja.

---

[1] http://dict.leo.org/?lp=ende&lang=de

Contents

List of Figures

List of Tables

Chapter 1

Introduction

Josef Reichle entered his workplace in a suburb of Baden-Baden in the morning of September 18th, 1973. When he switched on the light in the cooling chamber of the slaughterhouse, he and several of his collegues were immediately killed by an enourmous explosion that devastated the whole area [BB]. Because Josef Reichle happened to be a second cousin of mine, I was often told about this incident during my childhood as a tragic accident that could not have been avoided. In fact, the explosion was caused by very unfortunate circumstances. Natural gas escaped from a leaking cast-iron gate of a pipeline under the slaughterhouse, accumulated in the cooling chamber and, together with oxygen, formed a highly explosive mixture. Similar accidents could be avoided by using fixed installations of electrochemical gas sensors that are able to monitor a target gas and raise the alarm if the concentration exceeds a certain threshold. The situation that caused the disaster in Baden-Baden, however, could hardly have been foreseen because the gas pipeline was not directly connected to the cooling chamber. Therefore, an appropriate gas detection system would probably not have been installed even if it had been available then. Moreover, it is generally not feasible to install gas sensors at every place that is possibly endangered.

In the future when robots will be part of our daily lives in the domestic environment and at the workplace, surveillance of the ambient gas concentration could be performed by mobile robots that are equipped with an artificial sense of smell. This is especially desirable in the case of gases that cannot be sensed by humans. Carbon monoxide, for example, is still responsible for a large percentage of the accidental poisonings and deaths reported throughout the world each year [Pen99]. Carbon monoxide poisoning can be caused by a fire or inadequate ventilation and obstructed stoves, as is assumed to be the cause of Emile Zola's death 1902. There are sensors available nowadays that can compensate for our lack of sensitivity to this gas. In combination with the mobility provided by coming service robots, security robots or even entertainment robots, this presents the opportunity to monitor pollutants in a large percentage of our living space.

It is not required that gas surveillance tasks will be performed by dedicated inspection robots. While it is less likely that personal robots will be purchased solely because of the security aspect, it is to be expected that the capability to monitor the ambient pollutant concentration will improve the value of robots that are sold as a companion that monitors the health of his owners. It is also possible that gas sensitive devices and the required software will be offered as an inexpensive upgrade.

This thesis is concerned with the use of chemical gas sensors on a mobile robot. The possibility to measure the gas concentration with a mobile robot enables a broad range of applications. Aside from surveillance of environmental pollutants and the detection of hazardous gases, this also includes the use of self-produced odours to aid navigation [Eng89; Rus95] or to communicate with other robots, for example, by sending a chemical SOS signal [RKK00].

Apart from the detection of an increased gas concentration, the task of localising a distant source of gas is very important. In the case of the mentioned disaster in Baden-Baden, the workers in the slaughterhouse noticed an unusual smell several days before the explosion. It was, however, not possible to locate the cooling chamber as the place next to the source of gas. A considerable part of this thesis deals with the problem of gas source localisation in a natural environment. Providing robots with this ability is very promising for applications such as automatic humanitarian demining [RK03] or localisation of the victims of an avalanche. A further possibility would be an "electronic watchman" that is able to indicate and locate dangerous gas leaks, leaking solvents or a fire at its initial stage. Gas source localisation, however, is an intricate task due to the turbulent nature of gas transport under natural conditions, which leads to a patchy, quickly fluctuating gas distribution where the actual gas source is usually not located at the point of highest concentration. In addition to the interest in solutions of this problem due to possible applications, investigating gas source localisation strategies might also lead to a deeper understanding of the physical properties of turbulent motion, as well as the way in which animals use odours for navigation purposes.

Under the assumption of an enclosed 2D area and a sufficiently strong constant airflow, the gas source localisation problem can be broken down into three subtasks [HMR01]:

- *plume finding*: detecting an increased concentration,

- *plume traversal*: following the gas plume to its origin,

- *source declaration*: determining the certainty that the source has been found.

Although the existence of a constant plume is not guaranteed in an unventilated indoor environment, this classification can be applied without the assumption of a strong constant airflow in a similar way [LRZ03]:

- *gas finding*: detecting an increased concentration,

- *source tracing*: following the cues determined from the sensed gas distribution (and eventually using other sensor modalities) towards the source,

- *source declaration*: determining the certainty that the source has been found.

The task of gas finding is not addressed in this work. It can be accomplished by checking whether the gas concentration exceeds a suitable threshold during exploration of the inspected area. This thesis is rather concerned with the latter two issues. Apart from a statistical investigation of different gas source tracing strategies, it presents an experimental investigation of gas source declaration, an issue that has been paid little attention so far. Because the absolute concentration maximum is often found far from a gas source, the problem is to find regularities (both spatial and temporal) in a turbulent distribution that enable classification of an object depending on whether it is a source of gas or not. A gas source declaration method is required in addition to gas source tracing strategies to provide a complete solution to the gas source localisation problem. Moreover, the classification capability of gas source declaration itself is of potential use even if the full gas source localisation problem cannot be accomplished using only a sense of smell. An object that is to be classified could be located using other sensor modalities, and attributed based on gas sensor measurements. Possible applications include the identification of suspicious items containing illegal narcotics or explosive materials and employment in rescue robots to determine whether a victim is alive by assessing whether that person is a source of carbon dioxide.

A third main issue that is addressed in this work is gas concentration mapping. Here, it is important to take into consideration the peculiarities of chemical gas sensors and gas transport. Because of the temporally fluctuating gas distribution and the fact that gas sensor measurements provide information about a very small area, it is not possible to measure the entire concentration field at the same time without using a dense grid of sensors. On the other hand, it is often sufficient to know the time-constant structure of a gas distribution. This might be more important than knowledge about the location of a gas source in some cases because it can provide information about the question of where a high concentration is to be expected. Thus, mobile robots that inspect a contaminated area could be used in rescue missions in order to provide incident planning staff with information to prevent rescue workers from being harmed or killed due to explosions, asphyxiation or toxication [MCH+00]. An algorithm to create a gridmap representation of the average concentration field from sparse point samples of the instantaneous distribution is introduced in this work. A concentration gridmap can be build with this algorithm from the measurements recorded along the path of the robot, while the path not necessarily has to cover the inspected area uniformly.

1.1 Contributions to Mobile Robotics

This thesis is concerned with gas sensing on a mobile robot, especially the problems of gas distribution mapping, gas source tracing and gas source declaration. In addition to investigations in this field, a few contributions to mobile robotics are also presented in this work, which arose as a byproduct of the main research direction. The specific contributions include:

1. The introduction of a software architecture (especially designed but not restricted to robot control software) that allows the designer to map the functional units of a program to objects and to model the cooperation between these objects and the timing by dynamically configurable data flow chains.

2. Development of an inexpensive and quick to install absolute positioning system, which uses a number of web-cameras to track the 2D coordinates of a distinctly coloured object with centimetre-level accuracy.

3. Development of the Mark III mobile nose, an inexpensive gas-sensitive system that is intended to be used on a mobile robot, and the introduction of a method to determine the response characteristics of a mobile nose.

4. Investigations of different gas source localisation strategies without the assumption of a strong constant airflow, including a detailed statistical analysis based on a large number of trials.

5. Experimental observation of the effect that a strong correlation between sensor response and proximity to a source can be obtained in certain scenarios only if the robot is driven with a constant, sufficiently high speed.

6. Introduction of an algorithm to build a gridmap representation of the concentration distribution of a detected gas from multiple gas sensor measurements distributed over space and time. To the author's knowledge, the proposed algorithm provides the first solution to this problem suggested so far.

7. An experimental evaluation of different reactive gas source tracing strategies that apply a direct sensor-motor coupling, including a statistical analysis of the tracing performance.

8. Introduction and experimental evaluation of a modified gas source tracing strategy of the silkworm moth that does not depend on information about the local wind speed.

9. A proposal for a gas source localisation strategy (including gas source declaration) based on exploration and concentration peak avoidance. The main contribution is to point out a feature that can be used to identify a gas source

more reliably than from a maximum in the instantaneous concentration field, exploiting the fact that local concentration maxima occur more frequently near a gas source compared to regions distant from the source. The frequency of local maxima has not been considered for gas source localisation so far and the proposed strategy provides the first solution to the full problem of gas source localisation that is applicable if the source is not detectable as an obstacle.

10. Introduction and experimental evaluation of a method for gas source declaration that uses machine learning techniques to determine whether a gas source is located in the immediate vicinity of the robot from a series of gas sensor readings. To the best of the author's knowledge, this problem is explicitly addressed in this work for the first time.

1.2 Thesis Outline

The remainder of the thesis is organised as follows:

Chapter 2 gives an overview of gas sensing in the field of mobile robotics, including a description of the gas sensors used and a discussion of the peculiarities of machine olfaction in a natural environment, especially the random nature of turbulent gas distribution. Further, a review of related work on gas sensing with a mobile robot is provided.

Chapter 3 describes the hardware set-up of the two gas-sensitive robots that were utilised for the investigations in this work, and introduces a method to determine the response characteristics of a mobile nose.

Chapter 4 presents initial experiments in an indoor environment without ventilation, which were carried out to study the possibility of a direct correlation between the location of a gas source and the response of the gas sensors.

Chapter 5 introduces an algorithm for building concentration gridmaps and presents a discussion of experimental results regarding different aspects of the mapping process. A few required derivations are carried out in Appendix C, and the concentration gridmaps for all of the experiments are given in Appendix D.

Chapter 6 presents an experimental comparison of two reactive gas source tracing strategies that use a direct sensor-motor coupling, including a statistical analysis of the tracing performance. The path of the robot in all the trials is shown in Appendix E.

Chapter 7 introduces a modification of the gas source tracing strategy of the silkworm moth for use in environments without a strong airflow and presents an experimental evaluation of its performance.

Chapter 8 introduces a method for gas source declaration and presents an evaluation of the classification performance that can be achieved depending on the required level of accuracy.

Chapter 9 provides conclusions, a discussion of proposed solutions for the full gas source localisation task with respect to different scenarios, and an outlook on future work.

Appendix A describes the software architecture DDFLat, which allows the designer to map the functional units of an application to objects and to model the cooperation between these objects by dynamically configurable data flow chains. The DDFLat framework was used to implement all programs needed for this thesis.

Appendix B introduces the inexpensive vision-based positioning system W-CAPS, which was used for several experiments presented in this thesis. The system combines triangulation estimates obtained from a number of web-cameras to track a distinctly coloured object with centimetre-level accuracy.

1.3   Technical Remarks

Most of the figures in this thesis are colour-coded in order to enhance lucidity. However, all attempts were made to achieve an illustration that is also clear in the case of a black and white printout. When details of the figures are referenced in the text, this is made in a way that is applicable to either version of this thesis. Additional indications on a specific element of a figure that refer to the colour of the element are given in curly brackets. For example: "the main thread of this work is indicated in the text {in red}".

Colour versions of the figures in this thesis are available on-line at

    http://www.lilienthals.de/achim/research/

Chapter 2

Gas Sensing With Mobile Robots

"When I die and go to heaven, there are two matters on which I hope enlightenment. One is quantum electro-dynamics and the other is turbulence of fluids. About the former, I am really rather optimistic." (Horace Lamb [Vog97])

2.1 Electronic Nose and Gas Sensors

During the 1980s research on machine olfaction boomed [PD82; IK85; KIAI87; CK90] leading to a generally accepted definition of an electronic nose [Gar87; GB94] as an instrument that comprises an array of heterogeneous electrochemical gas sensors with partial specifity and a pattern recognition system [GB99]. Probably the first report of a similar device was published by Moncrieff at the beginning of the 1960s [Mon61]. Gas sensors are devices that measure the ambient gas atmosphere. A variety of different sensor types have been developed beginning with early work by Hartman, Wilkens and Sauerbrey [Har54; Sau59; WH64]. The gas detection principle is generally based on the fact that changes in the gaseous atmosphere alter the sensor properties in a characteristic way. Three main types of gas sensors have been used in connection with mobile robots, discussed as follows.

Acoustic Wave Gas Sensors

Acoustic wave gas sensors comprise a piezoelectronic substrate, usually quartz, and a coating with a specific affinity. By using different coatings the device can be made responsive to different gases. During operation, an alternating electric field is applied to generate an elastic wave in the quartz crystal. Temporarily absorbed molecules perturb the propagation of the acoustic waves due to the effect of the added mass and by changing the viscoelastic properties of the coating layer. The resulting shift of the fundamental frequency of the quartz crystal is then measured as the output of the sensor. A more detailed description is given by Gardner and Bartlett in [GB94].

Acoustic wave gas sensors are also known as quartz crystal microbalance (QMB or QCM) because the device can be regarded as a balance that is highly sensitive to the weight of gas molecules. Depending on whether the effect of surface waves or bulk waves is utilised, these sensors are also referred to as SAW or BAW devices. Advantages of the QMB sensor technology are the low power consumption, the possibility to control the selectivity over a wide range, long term stability and a long lifetime. On the other hand, QMB sensors exhibit a comparatively low sensitivity to the target gas and have a limited robustness to variations in humidity.

A general demand on gas sensors, especially concerning application on a mobile system, is a rapid response. The sensors should be able to react to and recover from an exposed gas within an acceptable time frame. Generally, the response characteristics depend not only on the type of sensor but also on the chemically sensitive layer used. Having this in mind, it can be said that acoustic wave gas sensors can offer rapid response. In particular, the time required for recovery is usually shorter compared to metal oxide gas sensors.

Metal Oxide Gas Sensors (MOX)

Metal oxide gas sensors are composed of a heating element inside a ceramic support tube coated with a semiconductor, most typically sintered tin dioxide (SnO_2). Gas is sensed by its effect on the electrical resistance of the semiconductor, which decreases in the presence of reducing gases. MOX sensors are operated at temperatures between 300°C and 550°C. A voltage across the heated surface causes an electrical current through the grain boundaries of the sintered polycrystalline surface of the semiconductor. Absorption of oxygen at the sensor surface increases the potential barrier between the grain boundaries, which causes a large effect on the sensor's resistance. The conductivity of the device thus sensitively reflects the rate of redox reactions with the ambient gas. A more extensive overview is provided in [GB94].

As a consequence of the high operating temperature, MOX sensors consume comparatively much power. In addition, the sensors typically have to be heated for 30 to 60 minutes before they can be used. Other disadvantages are the slow recovery after the target gas is removed, the weak durability over a prolonged time [Mor95] and the poor selectivity. Although it is possible to a certain extent to control the sensitivity for a particular target gas by using different sensor preparation methods and a varying operation temperature [Kap01], the combustion process is generally not strongly selective to the precise structural details of the gas molecules. The advantages that have made MOX sensors the most widely used gas sensor in electronic noses as well as in robotics are their high sensitivity (down to the sub-ppm level for some gases), the usable life-span of three to five years, and the comparatively low susceptibility to changing environmental conditions.

Conducting Polymer Gas Sensors

As in the case of MOX sensors, the measurand of conducting polymer sensors is the resistance of the surface layer. Instead of a semiconductor, a thin polymer film is used, which is usually deposited across the gap between two gold electrodes by electrochemical polymerisation. Typical polymers are constructed from monomers such as pyrrole, aniline or thiophene. In contrast to MOX sensors, the DC electrical resistance is monitored at room temperature. Conducting polymers respond to a wide range of organic vapours. The exact mechanism by which the conductivity is altered, however, remains unclear. Several interactions that might account for the observed effect are discussed by Gardner and Bartlett in [GB94].

Conducting polymer sensors are comparatively easy to prepare (although the conditions have to be controlled very carefully and the chemicals have to be suitably purified in order to achieve reproducible results) and a wide range of materials with a varying sensitivity for different organic gases can be synthesised. Because the sensors are operated at room temperature, i.e., no heating is required, the power consumption is low. The response to a target gas is typically fast with rapid recovery when the gas is removed. On the other hand, the actual level of sensitivity is approximately one order of magnitude lower than that of metal oxide sensors. Further disadvantages are the poor scientific understanding, the sensitivity to humidity and long-term drifts, which make conducting polymer sensors difficult to handle.

2.2 Gas Sensing in a Natural Environment

Electronic noses have been extensively studied under laboratory conditions. Numerous publications are available, for example, in the field of food analysis. This includes tests on the freshness of fish [OMO+92], quality estimation of ground meat [WHSL93], recognition of illegally produced spirituous beverages [KLZV99] and the discrimination of different coffee brands [HWRG99].

These remarkable results cannot be obtained in the same way on a mobile robot due to the influence of varying environmental conditions and restrictions because of limited resources such as power and the available space, for example. An important aspect that usually cannot be transferred to a mobile system is the sample handling process. In most laboratory-based applications much effort is expended to prepare the volatile components before they are analysed with the gas sensor system. Typically headspace samplers are used, which prepare a defined sample of evaporated material (the so-called headspace) and deliver the sample in a well defined way to a chamber that contains the gas sensors [KE97]. Moreover, best performance was achieved in laboratory experiments by using a measurement technique that requires a second gas (e.g. clean air with known humidity), which is periodically routed to the sensor chamber. It serves as the carrier gas for the headspace sample and as a reference for tracking the baseline level of the sensor response. Due to the demand

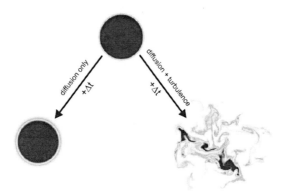

Figure 2.1: *The effect of turbulence compared to mere molecular diffusion, taken from Smyth and Moum [SM01]. At the top, the assumed initial state is depicted (a circular region of nearly homogeneous concentration) while the pictures on the bottom show two numerical solutions of the equations of motion obtained in the case of a motionless fluid (left) and in the case of fully developed, two-dimensional turbulence (right). The distribution shown on the right exhibits no smooth concentration gradients that indicate the direction toward the centre of the gas source.*

for real-time operation and restrictions of weight, space and power consumption it is usually not feasible to establish the same sample handling process on a mobile robot.

As a consequence of the limited resources it is also difficult to achieve sufficiently stable environmental conditions. Gardner and Bartlett report typical accuracies of $\pm 0.1°$C temperature, $\pm 1\%$ relative humidity and $\pm 1\%$ flow rate as important conditions when employing electronic nose technology [GB99]. For application on a mobile robot, larger variations in the environmental conditions have to be tolerated. Consequently, research in this field has focussed so far on using gas sensors for detection of a known target gas and localisation of the gas source rather than discrimination of different gases. To avoid confusion, a complete gas-sensitive system used on a mobile robot is sometimes referred to as a "mobile nose".

Gas Distribution in a Natural Environment

A major problem for gas source localisation in a natural environment is the strong influence of turbulence on the distribution of gas. Turbulent transport generally dominates the dispersal of gas due to molecular diffusion. For example, the diffusion constant of gaseous ethanol at $25°$C and 1 atm is D = 0.119 cm^2/s, corresponding to a diffusion velocity of 20.7 cm/h [NIM99]. Apart from very small distances where

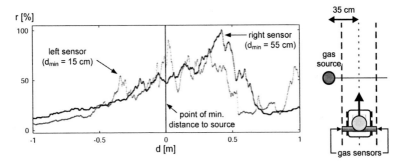

Figure 2.2: *Example of gas sensor readings recorded while the robot passed a gas source (ethanol) along a straight line (as shown on the right side) at a low speed of 0.25 cm/s. The curve displays relative conductance values of two metal oxide sensors mounted on the left and right side of the robot with a separation of 40 cm.*

turbulence is not effective, molecular diffusion can thus be neglected concerning the spread of gas.

Turbulent flows exhibit several general characteristics [RW02]. First, this includes their unpredictability. Turbulence is chaotic in the sense that the instantaneous velocity (and consequently also the instantaneous concentration of a target gas) at some instant of time is insufficient to predict the velocity a short time later. Second, the turbulent transport is considerably faster than molecular diffusion. This is indicated in Fig. 2.1, which shows simulated gas mixing due to the effect of diffusion alone (bottom, left) and due to diffusion and turbulence (bottom, right). The pictures in the lower part of the figure show snapshots of the gas distribution that evolved from the the circular distribution depicted in the upper part. These distributions were obtained by means of a numerical solution of the equations of motion by Smyth and Moum [SM01]. It is apparent from the figure that turbulent flow causes a much quicker spreading of the target gas. As an average effect this can be modelled by defining an effective turbulent diffusion coefficient (the eddy diffusivity, see for example [Hin75]). Third, a turbulent flow comprises at any instant a high degree of vortical motion. These continuously fluctuating eddies range in size from the largest geometrically bounded scales of the flow down to the scales where only molecular diffusion is effective. Small scale eddies stretch and twist the gas distribution, resulting in a complicated patchy structure (see Fig. 2.1). The instantaneous distribution exhibits no smooth concentration gradients that indicate the direction toward the centre of the gas source. Assuming a uniform and steady flow, however, the time-averaged concentration field varies smoothly in space with moderate concentration gradients. A fourth characteristic of turbulence is the dissipation of kinetic energy. Turbulent kinetic energy is passed down from the largest eddies to the smallest, where it is finally dissipated into heat by viscous forces (this is called the energy cascade). The magnitude of the viscous forces determines the minimal eddy size (Kolmogorov microscale, see [Rus99b; RW02] for details).

A further important transport mechanism for gases occurs due to the fluid flow itself (advective transport). This mechanism is effective even in an indoor environment without ventilation due to the fact that weak air currents exist as a result of pressure (draught) and temperature inhomogeneities (convection flow).

Ultimately, the concentration field of a target gas released from a static source is a fluctuating asymmetrically shifted distribution of intermittent patches of high concentration with steep gradients at their edges. These properties are illustrated in Fig. 2.2, which shows typical sensor readings in the vicinity of a gas source (evaporating liquid ethanol). In this experiment, the robot passed the source along a straight line at a low speed of 0.25 cm/s. Even though there is a smoothing effect due to the slow recovery of the metal oxide gas sensors used, the curve in Fig. 2.2 reveals the existence of many local concentration maxima. Furthermore, the absolute maximum is mostly located some distance from the actual location of the gas source due to the fact that the concentration in isolated gas patches does not depend strongly on the average concentration. In the case of turbulent diffusion from a small source, the peak concentration values are generally an order of magnitude higher compared to the time-averaged values [RW02].

2.3 Scientific Work in the Field of Gas Sensing With Mobile Robots

Chemical sensing entered the field of mobile robotics in the beginning of the 1990s. Early work focused on the use of gradient-following (chemo-tropotaxis) for gas source tracing without an explicit reference on the environmental conditions. Rozas, Morales and Vega report on their observation of a decreasing concentration with increasing distance to a gas source [RMV91]. This result was found by means of measurements with a ventilated gas capture device (containing six metal oxide sensors) that was mounted on a mobile robot. The measurements were performed at four different distances from the source (0 m, 0.5 m, 1 m and 3 m, respectively). Because only one trial is reported for each of the three analytes tested, the reproducibility of the results is unclear. Due to the turbulent character of gas propagation, however, a monotonic relationship between the observed concentration and the distance to the source cannot be expected in general.

Sandini et al. considered the problem of gas source tracing as an example of a localisation task, which has to be performed based on local information that does not directly indicate the source location [SLV93]. As a possible solution, the authors suggest a strategy that involves periods of random exploration and gradient following, and the robot switches between these states depending on whether the concentration is above or below a certain threshold. Gradient following was implemented by applying a positive or negative driving command depending on the sign of the concentration gradient sensed with a pair of metal oxide sensors. The authors point out that a possible limitation for a gradient-following strategy based on chemical sensor measurements is the fact that the instantaneous concentration gradient might not be accessible because of the implicit temporal integration that is performed by the sensors due to their long decay time. Quantitative results of their gas source tracing experiments are, however, not given and the environmental conditions are not specified in sufficient detail to permit meaningful comparison with similar experiments.

Classification of the Mobile Nose Literature

Subsequent work in the field of airborne chemical sensing with a mobile robot, which includes a description of the environmental conditions, is basically concerned with three problem domains:

A. *trail guidance,*

B. *localisation of a distant gas source,*

C. *gas distribution mapping.*

Concerning gas source localisation, the available publications can be further divided into the subtasks of gas finding, source tracing or source declaration, and also on whether a strong airflow is assumed or not. A strong airflow of at least 5 cm/s has to be assumed if information about the local wind vector is required by the control algorithm. In an industrial or domestic indoor environment with moderate ventilation, however, wind fields with velocities less than 5 cm/s are typically encountered [INM+01; PSM86]. According to this classification scheme, the following literature survey is organized as

- *gas distribution mapping,*

- *trail guidance* (low sensor-to-source distance),

- *suggestions for gas finding,*

- *gas source tracing in an indoor environment with a strong constant airflow* (information about the local wind vector is available),

- *tracing of a distant gas source in an indoor environment without a strong constant airflow* (information about the local wind vector is not available),

- *suggestions for gas source declaration.*

Publications that are based on simulations are not considered because a faithful model of gas distribution in a real world environment cannot be obtained with currently available simulation techniques.

This thesis is concerned with gas source localisation (especially the subtasks of gas source tracing and gas source declaration) without information about the local wind vector, i.e., localisation strategies that are not restricted to a strongly ventilated environment. In addition, the aspect of gas distribution mapping is addressed.

2.4   Gas Distribution Mapping

A straightforward method to create a representation of the time-averaged concentration field is to measure the response over a prolonged time with a grid of gas sensors. This technique has been used on various occasions by Ishida and co-workers. The time-averaged gas sensor response over 5 minutes at 33 grid points distributed over an area of 2×1 m^2 was used in [INM98], for example, to characterise the experimental environment. With an increasing area, however, establishing a dense grid of gas sensors would involve an arbitrarily high number of fixed gas sensors, which poses problems such as cost and a lack of flexibility. Furthermore, an array of metal oxide sensors (used by Ishida et al.) would cause a severe disturbance due to the convective flow created by the heaters built into these sensors [ITYM03].

Gas measurements acquired with a mobile robot were used by Hayes et al. to create a representation of the gas distribution by means of a two-dimensional histogram [HMG02]. The histogram bins contained the number of "odour hits" received in the corresponding area while a random walk behaviour was performed. "Odour hits" are counted whenever the sensed concentration exceeds a defined threshold. In addition to the dependency of the gas distribution map on the selected threshold, the problem with using only binary information from the gas sensors is that much useful information about fine gradations in the average concentration is discarded. It would also take a very long time to obtain statistically reliable results, and there is no extrapolation on the measurements apart from the quantisation into the histogram bins. So it is doubtful whether this approach would scale well to larger environments. A further disadvantage of this method is that it requires perfectly even coverage of the inspected area by the mobile robot.

2.5   Trail Guidance

Olfactory markings are often used by animals to store and communicate spatial information. A well-known example concerns ants that mark the path to a source of food with an odour trail [Sud67]. Because the information is stored physically in the environment, there is no need for an abstract representation of the environment in the animal's brain. The odour trail is refreshed continuously by ants that follow the path. By varying the intensity and the frequency of trail marking, the trail is

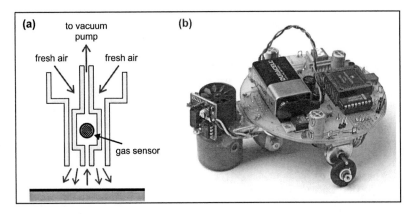

Figure 2.3: *Air curtain technique. (a) Illustration of the functional principle, adapted from Russell et al. [RTMS94]. Air is drawn from the floor through the sensor inlet and blown in the opposite direction around the sensor inlet. Thus, an outward airflow is created that is able to deflect gas carried towards the sensor by external air currents. (b) Image of the mobile robot "Nose-Bot" equipped with the improved gas sensor (photograph by courtesy of R. Andrew Russell).*

also used to communicate the quality of the food it leads to. Chemical markings can also contain temporal information due to their naturally fading intensity. Honey bees, for example, use odourous markers to increase their efficiency when gathering nectar. After visiting a flower and gathering its nectar, the honey bee marks the flower with a short-lived odour, which provides a warning that the flower has not had time to make more nectar [GN92].

Using chemical markings is of possible benefit for a number of applications in the field of mobile robotics. Olfactory trails could provide an inexpensive and more flexible alternative to the metal wires buried under the floor that are often used for industrial automated guided vehicles (AGV) [SMVD95]. Apart from establishing a path to follow, odourous trails could also be used in order to provide a temporary repellent marking, which indicates areas on the floor that have been cleaned, for example [Eng89]. While this would be particularly beneficial to coordinate the behaviour of multiple robots, it could also be helpful in the case of a single robot, because it avoids the necessity for maintaining a consistent spatial representation. Further application scenarios of trail guidance for mobile robotics are discussed by Russell in [Rus95].

In contrast to the task of localisation of a distant gas source, the impact of turbulence is considerably reduced in the case of trail guidance because of the low sensor-to-source distance, which is in the order of 1 cm in most of the experimental work published in this area so far. Odour trails placed on the floor are covered by a layer where the airflow is laminar. This layer is so thin that current robots cannot measure concentrations in this region [Rus99a]. However, compared with experiments in which a distant source is to be localised, the proximity of the sensors to the trail causes a much stronger differentiation of the average concentration gradient in the sensor signal. Furthermore it is possible to increase the differentiation near the floor by introducing well adjusted additional airflows to block external ones, thus establishing an "air curtain" [RTMS94] as indicated in Fig. 2.3.

Several navigation strategies have been suggested for trail following that assume a pair of gas sensors, which samples the analyte concentration closely above the ground. A possible method to follow a broad trail (wider than the sensor spacing) was introduced by Stella et al. [SMVD95]. It is based on the idea of rotating the robot back if one sensor detects a considerably lower concentration of the analyte, thus trying to keep both sensors over the trail (see Fig. 2.4, a). With a pair of conductive polymer sensors (spacing: 10 cm, distance to the ground: 1 cm), their robot could successfully follow a 15 cm wide and 4 m long trail of alcohol with moderate turnings at a speed of 6 cm/s in the reported experiment.

A similar algorithm suggested by Russell et al. [Rus95] tries to follow a trail between the sensors by rotating the robot towards the higher concentration using a direct sensor motor coupling (see Fig. 2.4, b). Experiments were carried out on a Mars mobile robot [DTRMS94], using a pair of quartz microbalance sensors with a spacing of either 3 cm or 5 cm and a distance to the ground of approximately 0.5 cm. With either set-up the robot was able to follow a narrow, continuous camphor trail, comprising two straight sections with a length of 50 cm each and a 30 degree turn inbetween.

Further trail following strategies that are more robust against sensor errors and imperfections in the trail are inspired by the behaviour of ants, which includes frequent crossings of the trail along a sinusoidal walk [Han67]. Ants are able to follow a pheromone trail with one antenna removed and even with their antennae crossed over. While trail following performance might be improved by choosing an appropriate non-linear sensor motor transfer function for the turning back behaviour [SW98], it seems to be most promising to add a time-out mechanism that enables the robot to swing back towards the expected location of the trail when the robot fails to detect the trail for a certain amount of time (see Fig. 2.4, c). Experiments with a six-legged walking robot that applies such an ant-like strategy showed the feasibility of the approach in the case of an intermittent trail [Rus99a]. This result was obtained with a pair of special gas sensors, consisting of a quartz microbalance sensor and a device to generate an "air curtain" (see Fig. 2.3, a) in order to increase spatial differentiation of the sensor signal. The sensors were mounted 13 cm apart and approximately 1 cm above the ground. The legged robot could follow an approximately 2 m long camphor trail, consisting of a straight section, a slightly curved section and a large gap of approximately 40 cm.

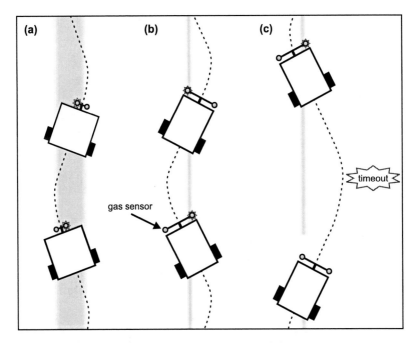

Figure 2.4: *Trail following strategies with a pair of sensors. (a) Chemo-tropotaxis to keep a pair of sensors over the trail. (b) Chemo-tropotaxis to keep the trail between the pair of sensors. (c) Timeout mechanism to increase robustness against measurement errors or in case of intermittent trails. The idealised odour trail is indicated by a bright grey vertical stripe, and a {red} star around a sensor indicates that this sensor detects the trail.*

A few implementations of chemical guidance on a mobile robot have also been published where the analyte concentration was sampled with a single gas sensing device (see Fig. 2.5). Larionova et al. [LAMdA03] report on their prototype of a cleaning robot, which uses a single array of metal oxide sensors and a system with small fans to force the air from the ground up while external airflows are blocked (similar to the "air curtain" technique). In preliminary tests it was demonstrated that the robot could detect a liquid chemical (10% alcohol mixture) laid on the floor by a human cleaner while the robot was driven with a speed of approximately 1.3 cm/s.

Further implementations address the task of following the edge of a trail with a single sensor. An algorithm to achieve this task was suggested by Russell et al. [Rus99b]. The procedure is shown schematically in Fig. 2.5 (b). Under the

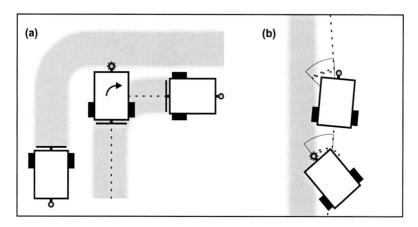

Figure 2.5: *Trail guidance strategies with a single gas sensor. (a) Repellent chemical marker. The example shows cleaning robots that use an odourous trail to indicate the area that is already cleaned. (b) Trail following with a single gas sensor [Rus99b].*

assumption that the robot is initially situated so that the sensor can be rotated over the trail, a three step edge tracking strategy is applied. First, the robot is turned until it detects the trail border. Then, it rotates a fixed angle away from the trail edge and moves a fixed distance forward. This sequence is repeated in order to follow the trail. Implementations of this algorithm were reported by Russell [Rus99b] and Mann and Katz [MK99]. In both cases a quartz microbalance sensor was used. With their implementation on a circular robot with a diameter of 10 cm, Russell et al. reported that a tracking speed of 1.7 cm/s could be achieved. The general feasibility of this approach was confirmed by Mann and Katz based on their floor cleaning tests with a prototype cleaning robot (footprint: 32×32 cm^2). However, a thorough analysis of the edge tracking performance achieved was not given by the authors.

2.6 Suggestions for Gas Finding

Possible strategies to make contact with the target gas are discussed by Russell et al. in [RBHSW03]. A solution that minimises energy consumption is *passive monitoring*, where the robot remains stationary until it detects an increased gas concentration. Patches of gas can be detected in this way even if the robot is located far away from the gas plume, due to the turbulence of airflow.

In order to accelerate the process of gas finding, the designated area has to be explored actively. By modelling chemical plumes as a straight line with a limited

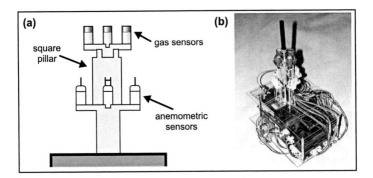

Figure 2.6: *Sensor probe used in the experiments of Ishida et al. [ISNM94], which consists of four thermistor anemometric sensors and four metal oxide gas sensors. (a) Sketch of the probe, adapted from [ISNM94]. (b) Mobile robot carrying the sensor probe (photograph by courtesy of Hiroshi Ishida).*

length, it can be shown that it is beneficial to carry out exploration along straight paths orthogonal to the wind direction [RBHSW03]. If information about the direction of the wind flow is not available, gas finding can be considered as a basic search task. Depending on the sensor equipment of the mobile robot, searching might be accomplished by a simple random walk behaviour or more sophisticated exploration strategies. It has to be noted, however, that searching for patches of gas is not guaranteed to succeed even if the search path covers the whole inspected area due to the temporal variation of a turbulent gas distribution [HMG02]. An experimental comparison of different gas finding strategies, however, has not been published so far, to the best knowledge of the author.

2.7 Gas Source Tracing in an Indoor Environment With a Strong Constant Airflow

Chemical trails marked on the ground form a relatively stable concentration profile in the vicinity of the trail. The signal processing part of the problem of trail guidance is the detection of the transition between a chemically marked area and its unmarked vicinity from sensor readings obtained at a low distance from the ground. In the case of gas source tracing, the task is to extract information about the location of a distant gas source from local concentration measurements sampled from a turbulent gas distribution. Because chemical stimuli are not inherently directional, this information has to be derived from at least two spatially or temporally distributed samples. When reliable information about the air flow is available, the local upwind direction can additionally be used as an indication of the direction to the source.

Most experiments with gas source tracing robots assume a strong unidirectional airflow that enables two step strategies, which combine gradient-following (tropotaxis) and periods of upwind movement (anemotaxis). Ishida et al. [ISNM94] introduced a remotely controlled mobile platform equipped with a probe consisting of four thermistor anemometric sensors and four metal oxide gas sensors (TGS 822), which is shown in Fig. 2.6. The wind sensors were mounted around a square pillar with a spacing of 90°, thus making it possible to obtain information about the direction of the airflow. Ideally, the direction of the sensor with the lowest output should correspond to the downwind direction since the wind is obstructed by the pillar. The gas sensors were mounted on top of the pillar, each located straight above one anemometric sensor. With this experimental platform, two different plume tracing methods were tested in a small wind tunnel (0.7 × 0.8 × 0.35 m³). The gas source was provided by a nozzle that spouted ethanol gas at a rate of 150 ml/s, and an average wind speed of approximately 20 cm/s was generated by a fan.

Step-by-Step Progress Method

The idea of the step-by-step progress method is to follow the concentration gradient towards the centre of a gas plume and to move upwind at the same time. Gradient-following is performed across the wind direction where the concentration gradient is usually higher than along the wind direction. The step-by-step progress method is shown in Fig. 2.7 (a). It was implemented as follows: first, the wind direction was measured and the readings of the two gas sensors perpendicular to this direction were compared. In the experiments, the wind direction was determined with an accuracy of 90° by selecting the anemometric sensor with the lowest output. The intermediate angle between wind direction and the gas sensor with the larger response was then chosen as an approximation to the direction of the gas source. Finally, the robot was rotated in this direction and driven a small distance (2 cm) forwards. This procedure was repeated until the robot reached the edge of the wind channel (the source was placed outside the wind channel). Because the concentration gradient across the wind direction was often found to be misleading with very low gas concentrations, a waiting period was introduced. The robot was stopped when both of the sensor readings used to determine the concentration gradient fell below a fixed threshold. With this additional mechanism, the step-by-step progress method was found to be successful under different conditions where the average wind velocity was reduced to 12 cm/s and/or a lower spouting rate of the gas source was used. The speed of the robot was set to 1.1 cm/s.

Zigzag Approach Method

The second method tested by Ishida et al. is based on the idea of crossing the plume repeatedly at an angle to the upwind direction until the edge of the plume is detected (zigzag approach method). It was implemented as follows: first, the robot is rotated to an angle of $\alpha = 60°$ with respect to the upwind direction and moves along a

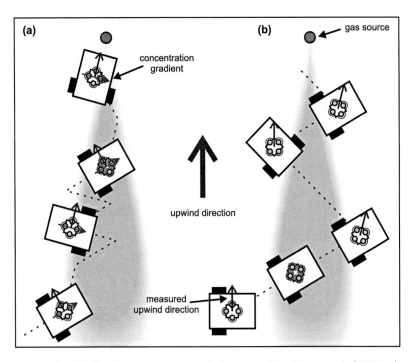

Figure 2.7: *(a) Step-by-step progress method suggested by Ishida et al. [ISNM94]. The wind direction was determined with an accuracy of 90° in this experiment. For clarity of the illustration, a larger step length (in proportion to the size of the robot) than in the actual experiment was assumed. (b) Zigzag approach method [ISNM94]. The wind direction was determined with an accuracy of 45° in this experiment. Possible trajectories are indicated on top of an idealised gas plume. A {red} star around a sensor indicates that this sensor detects the target gas.*

straight line until it detects the beginning of the plume. The robot carries on driving in a straight line until it detects the end of the plume, then it rotates back to an angle that is alternately set to $-\alpha$ and α with respect to the upwind direction. In the experiments, the upwind direction was determined with an accuracy of 45° from the response pattern of the anemometric sensors. The plume was detected by observing whether the gas sensor readings exceeded a fixed threshold. It was found, however, that it was necessary to add a mechanism to relocate the plume in cases where the robot moved out of the plume due to spurious turns caused by fluctuations in the sensor readings. This mechanism was implemented as a backtracking movement

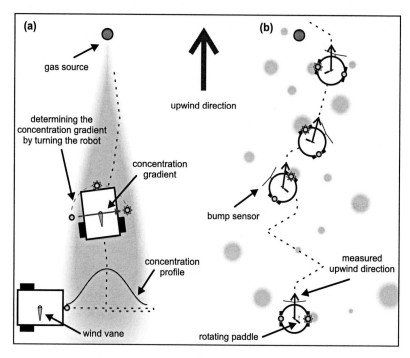

Figure 2.8: *(a) Plume-centred upwind search by Russell et al. [RTDMS95]. (b) Bombyx mori algorithm as implemented by Russell and co-workers in [RBHSW03]. Possible trajectories are plotted on top of the idealised gas distribution assumed by the particular strategy.*

that was triggered when the sensor readings fell below a fixed threshold. With this additional mechanism, the second algorithm was also successful in tracing the plume of the gas source in the sense that the robot reached the end of the wind channel approximately level with the source. Apart from a demonstration of the general feasibility of the approach under the condition of a strong unidirectional airflow, the experiments of Ishida indicate the importance of a mechanism to deal with erroneous decisions that occur due to the turbulence of the air flow.

Plume-Centred Upwind Search

A gas source tracing strategy for a robot with a single gas sensor and a wind measuring device was proposed by Russell et al. [RTDMS95]. It is sketched in Fig. 2.8 (a). The strategy was implemented on a wheeled robot that was equipped with a single

QMB sensor and a wind vane to measure the direction of the airflow. First, the robot tries to find the centre of the gas plume. This is achieved by recording the concentration profile while the robot moves across the wind direction. When the robot reaches the far side of the plume (indicated by a sensor reading that falls below a certain threshold), it returns to the calculated centre of the plume and turns into the upwind direction. Then, the plume tracing phase is started, which is similar to the step-by-step progress method discussed above. To maintain a path close to the centre of the plume, the robot corrects its heading after each step (the step length was 0.35 cm in the experiment) according to the concentration gradient across the wind direction, which is sensed by turning the robot 90° to the left and right, respectively. Russell and co-workers report two successful trials in a corridor with a unidirectional airflow of approximately 30 cm/s where people occasionally walk past the robot during the experiments. The robot was able to follow the gas plume over a distance of approximately one metre with and without an obstacle inbetween the gas source and the point where the tracing phase was started. In the experiments a strong gas source was assembled from a number of short sections of cardboard mailing tube glued together, which were then sprayed with a solution of camphor dissolved in alcohol immediately before each experiment. The camphor plume was created by a fan that generated an airstream through the open-ended tubes.

Biological Inspiration: Upwind Movement and Local Search

Further implementations draw inspiration from the observation of biological systems that apply a combination of anemotaxis and chemo-tropotaxis. A well investigated example is the gas source tracing behaviour of the silkworm moth *Bombyx mori*. Male moths are able to localise females that release a specific pheromone (Bombycol) over large distances. They use a rather simple behaviour to cope with the problem that the odour plume occurs as intermittent patches of high pheromone concentration due to the turbulence of the airflow. The behaviour consists of a programmed motion sequence that is (re-)started whenever a patch of pheromones is detected. The motion sequence carries out an oriented local search for the next pheromone patch. It consists of an initial forward surge in upwind direction, followed by a side-to-side search (performed with increasing amplitude) and a final looping motion. A detailed description is given in Section 7.2 and in [Kan96].

Kuwana et al. developed an experimental set-up that enables comparison of gas source tracing strategies implemented on a mobile robot with the tracing behaviour of a real moth under the same conditions [KNSK99]. In order to achieve a high degree of comparability, the behaviour was implemented on a small robot (with a similar size to the real moth) and living antennae taken from a moth were utilised as a pair of gas sensors [KSM95]. However, a comparison with a robot controlled by an algorithm that mimics the behaviour of the moth has not yet been published by Kuwana and co-workers. Rather, an experiment is reported with a robot controlled by a simple reflex-based program, which performs chemo-tropotaxis without using information about the local wind vector. The robot could trace a pheromone source over a distance of 10 cm in a wind tunnel with a wind speed of 25 cm/s. While a real moth could localise the pheromone source very precisely in this environment, the robotic moth missed the source in the reported trial by almost 2 cm and did not stop after passing the pheromone source [KNSK99].

Silkworm Moth (*Bombyx mori*) Algorithm

A gas source tracing strategy that actually mimics the behaviour of the moth *Bombyx mori* was implemented by Russell and co-workers [RBHSW03]. A mobile robot was equipped with two polymer gas sensors and a wind measuring device, consisting of a small rotating paddle and an optical encoder to measure the rotational speed of the paddle [RK00]. From the variation in the rotational speed of the paddle both the wind velocity and wind direction can be determined. The circular robot with a diameter of 10 cm had a similar construction to a Logo Turtle. It had two laterally mounted wheels and a third point of contact was provided by a teflon pad. The experimental environment was the top of a table tennis table (2.7 × 1.5 m^2). A fan that produced an airflow of approximately 1.5 m/s was situated 2.8 m from the point where the robot was started and approximately 1 metre from the gas source (a conical flask filled with 5% ammonia solution through which air was bubbled by a pump). The *Bombyx mori* behaviour was implemented in an iterative manner (see Fig. 2.8 (b)). First, the robot turns towards the wind direction and waits until it detects an increased gas concentration. Next, a local search is carried out, consisting of a forward step (10 cm) and a zigzag movement. The zigzag movement comprises a step towards the sensor that detected the target gas (±60° with respect to the wind direction), a double step (20 cm) towards the other sensor (∓60°) and a final step in the same direction as the first one (±60°). Then, the robot turns again towards the wind direction and moves in a circle (with a radius of 10 cm) in the direction of the sensor that was first stimulated. Finally, the robot moves along the same circle in the opposite direction. After each step, the local search can be restarted from the beginning if another patch of gas is detected. When receiving no stimulation during this motion sequence, the algorithm returns to the initial waiting phase.

Dung Beetle (*Geotrupes stercorarius*) Algorithm

As another biologically inspired algorithm, the gas source tracing behaviour of the dung beetle *G. stercorarius* was implemented by Russell and co-workers [RBHSW03] (see Fig. 2.9 (a)). This algorithm is similar to the zigzag approach method introduced above. The robot is driven diagonally across the plume with an angle of ±60° to the upwind direction until it detects the outer edge of the plume. Then it is turned to ∓60° and the procedure is repeated. As in the silkworm moth algorithm, the dung beetle behaviour was implemented in a stepwise way. The diagonal movement was divided into 50 cm steps. Upon detecting the outer edge of the plume, the

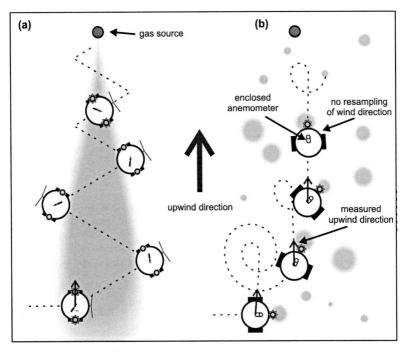

Figure 2.9: *(a) Dungbeetle G. stercorarius algorithm, Russell et al. [RBHSW03]. (b) Spiral surge algorithm, Hayes et al. [HMG02].*

orientation was changed at the end of a step. Otherwise, the next step was oriented in the same direction as the previous one.

Escherichia coli Algorithm

Two further gas source tracing strategies were tested by Russell and co-workers that do not use information about the wind direction [RBHSW03]. The *E. coli* algorithm consists of straight "runs" and direction randomising "tumbles". It was implemented as a sequence of straight steps after which either a small rotation (randomly chosen from $[-5°,5°]$) is performed if the current sensor reading is higher than the previous one, or the robot is turned by a larger angle (randomly chosen from $[-180°,180°]$) in the case of decreasing stimulation. Thus, the runs tend to be straighter if the sensed concentration increases while a high "tumbling frequency" is applied otherwise. However, Russell et al. report that a convincing gas source tracing behaviour could not be achieved with the E. coli algorithm.

Iterative Chemo-Tropotaxis

In contrast to the E. coli algorithm, which requires only a single gas sensor, the fourth strategy tested with the table tennis experimental set-up was an iterative version of reactive gradient-based control (iterative chemo-tropotaxis). It was implemented as a sequence of two repeatedly executed steps. The robot moves forward for a set distance (2 cm) and then rotates by an angle proportional to the sensed concentration gradient towards the side that was stimulated more. The maximum turn angle per step was set to $\pm 16°$.

Results of the Table Tennis Experiments

Due to the high wind speed, the concentration gradient along the gas plume was rather small. As a consequence, it was found that the iterative chemo-tropotactic strategy could not distinguish whether the robot was moving towards or away from the gas source. Thus, the algorithm failed in all of the ten trials where the robot was started facing in downwind direction. In contrast, five out of ten trials where the robot was started in upwind direction were successful in the sense that the robot finally collided with the gas source. A higher success rate of 70 % was found in ten trials with the silkworm moth behaviour where the robot was started from the same position as in the experiments with the iterative chemo-tropotactic strategy. The same success rate of 70 % was also observed in another ten trials where the dung beetle strategy was applied. The observed mean path length of successive trials was found to be higher for the dung beetle algorithm ($\approx$ 190% of the shortest path) compared to the silkworm moth algorithm ($\approx$ 130%) and iterative chemo-tropotaxis ($\approx$ 105%). It is, however, difficult to compare these results because of the small number of trials and the different starting positions. The silkworm moth algorithm, for example, produced a relatively short path because the local search was usually restarted before the looping phase. A longer relative path length could be expected if the robot was started farther away from the source where patches of gas occur more sparsely. Also the success rate should vary with the tracing distance. While the gradient-based tracing method, for example, was able to quickly acquire and follow the centre-line of the gas plume in the successful trials, it was also more susceptible to failures because it contains no means of recovering from erroneous situations. These situations will occur more frequently at a larger distance from the source where the peak to time-average concentration ratio is higher.

Spiral Surge Algorithm

The Spiral Surge Algorithm introduced by Hayes, Martinoli and Goodman [HMG02] is based on the same mechanisms as the behaviour of the silkworm moth. When an "odour packet" is detected, the robot determines the wind direction and moves upwind for a set distance (see Fig. 2.9 (b)). If another odour packet is encountered during the surge, the robot resets the surge distance but does not resample the wind direction. Next, an outward spiral is performed as a local search for the next "odour hit". This algorithm was implemented on a group of circular-shaped robots (diameter: 24 cm), which were equipped with a hot wire anemometer and a single conducting polymer gas sensor tuned to sense water vapour. The anemometer was enclosed in a tube to enable determination of the wind direction as the direction where the highest wind speed is measured while the robot performs a slow 360° rotation. While determining the wind direction is a rather time-consuming operation due to the involved scanning movement, the fast response time of the polymer gas sensor enabled a comparatively high speed during the searching behaviour (approximately 30 cm/s). A gas plume was created by a pan of hot water located in one corner of the test arena in front of a bank containing five fans, which produced an airflow of approximately 1 m/s. Remarkably, only binary information from a single gas sensor was used, based on whether a reading exceeds a fixed threshold or not. Consequently, there was no preferred direction to start the spiral search. The plume tracing performance was tested in a 6.7 × 6.7 m² arena. A successful trial was assumed when the robot entered a radius of 88 cm around the source.

Three variations of the spiral surge algorithm were tested. First, the Spiral Surge Algorithm with a long step size for the initial surge and a very large spiral gap (SS1), meaning that no local-search is performed because the search paths are straight. Second, the Spiral Surge Algorithm with a local search due to a small spiral gap of approximately 35 cm (SS2). And third, a modified SS2 behaviour that represents blind upwind searching because it receives "odour hits" without correlation to the position of the robot (random odour behaviour). The "odour hits" were internally generated from the sequence that was obtained in the SS2 experiments. 15 trials with each algorithm were conducted for each group size (one to six robots).

The experiments of Hayes et al. demonstrated the suitabililty of the Spiral Surge Algorithm for gas source tracing over a distance of more than 6 m. With high statistical significance the SS2 algorithm was found to perform better in terms of the distance travelled compared to the "random odour behaviour", indicating that the location where the spiral movement is started is in fact an important aspect of the searching strategy. The shortest tracing paths were obtained with the SS1 strategy. This is interpreted by the authors as an effect of the small test arena, where a local search for the next "odour hit" is not required because a straight surge in upwind direction usually guides the robot towards the source. With increasing group size, the random walk behaviour tends to approach the performance of the SS1 strategy, indicating that a gas source can be best found by chance in an overcrowded area.

Multiphase Tracing Algorithm

All gas source tracing strategies discussed in this section so far were tested in an environment with a strong and uniform airflow. In cases where the wind direction

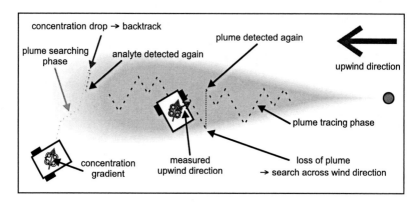

Figure 2.10: *Multiphase algorithm (Ishida et al. [IKNM96]).*

is not uniform, anemotactic search can be easily mislead by the unstable wind field
in regions where different air currents mix together. Ishida et al. report that their
gas source tracing algorithms, which were found to be successful in wind tunnel
experiments with uniform wind (see "Step-by-Step Progress Method" and "Zigzag
Approach Method"), were not always useful in experiments in a clean room with
two air supply openings [IKNM96]. The experimental area was a 2.8×1.0 m^2
region near to two air supply openings. In this area, the wind speed was in the
range of $10 - 30$ cm/s. The same experimental platform as in the wind tunnel
experiments was used, consisting of a remotely controlled mobile system equipped
with the gas- and wind-sensitive probe shown in Fig. 2.6. To provide a gas source,
a nozzle spouting saturated ethanol gas was situated near one air supply opening.

In order to cope with the non-uniform wind direction, Ishida and co-workers
proposed a multiphase tracing algorithm [IKNM96] as indicated in Fig. 2.10. The
fundamental element is a combination of upwind search and gradient-following across
the wind direction similar to the step-by-step progress algorithm introduced above.
Because the robot often failed to pass the intermediate region between the two
air supply openings when started from the inlet distant from the gas source, the
combination of anemotaxis and chemotaxis was applied only if the sensor readings
exceeded a certain threshold indicating a position near to the centre of the gas
plume (plume tracing phase). As long as the readings were below this threshold, the
robot was driven along the concentration gradient disregarding the wind information
(plume searching phase). With an additional mechanism to lower the threshold
when no considerable progress was made in the plume searching phase, the robot
could trace the gas source through regions of low concentration and unstable wind
direction by switching between the plume searching and plume tracing phase. The
proposed multiphase algorithm comprises also a backtracking mechanism to handle

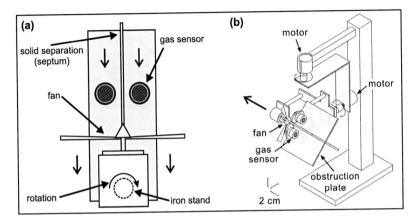

Figure 2.11: *(a) Sketch that shows the functional principle of the two dimensional odour compass (top view), adapted from [IM03]. (b) Three dimensional odour compass (by courtesy of Hiroshi Ishida).*

situations where a sudden drop of the measured gas concentration occurs due to the turbulence of the airflow.

While unsuccessful trials are reported using only pure chemotaxis (as applied in the plume searching phase) or exclusively using the combination of anemotaxis and chemotaxis (plume tracing phase), the multiphase algorithm was found to be able to overcome the difficulties of a non-uniform wind direction field in the clean room. The gas source could be traced over a distance of approximately 1.5 m using either $\alpha = 60°$, or $\alpha = 30°$. A successful trial is also reported with an alternative switching mechanism that does not require a threshold value. The change between the plume searching and the plume tracing phase was triggered in this case when not enough progress was made in one phase [IM03].

Odour Compass

In connection with the multitude of tracing strategies, another device has to be mentioned that was devised to support the gas source tracing process in the same way as a magnetic compass supports travelling towards the north pole. A sketch of the two-dimensional "odour compass" developed by Nakamoto et al. [NIM96] is shown in Fig. 2.11 (a). It consists of two metal oxide gas sensors and a fan situated on a support, which is mounted on a rotatable stand. The gas sensors are separated by a solid plate. An estimate of the direction of the local concentration gradient is obtained from the position where the two sensor responses match while the "odour compass" is rotated through 360°. The plate to separate the metal oxide gas sensors and the fan is used to enhance discrimination with respect to the orientation relative to a gas plume. Both mechanisms are biologically inspired. A solid separation (septum) is often found in biological noses indicating that the design has been an evolutionary success. Modulating the reception of chemical signals, for example by sniffing, is also a widespread behaviour in the animal kingdom. The use of a fan in the odour compass was inspired by the observation that wing-fanning is important for the gas tracing performance obtained by walking silkworm moths [NIM96]. Mimicking the effect of wing-fanning, the fan draws the target gas towards the gas sensors if a gas plume is situated in front of the odour compass, and pushes the target gas away from the sensors when the compass head is turned away from the plume. The best discrimination performance was achieved when the fans produced a wind speed of approximately 50 cm/s.

The concept of the odour compass was extended to a three-dimensional version (see Fig. 2.11 (b)) that uses four metal oxide gas sensors on a head, which can be rotated about two degrees of freedom. Although the estimated direction at one position exhibits considerable fluctuations of up to 30° both in horizontal and vertical direction, a gas source could be traced by repeated steps along the indicated direction. This was demonstrated by Ishida and co-workers in the same clean room where the multiphase algorithm was also tested. The odour compass was moved manually and the step size was either 20 cm or 30 cm. Successful three-dimensional tracing of an ethanol source (75 ml/min) over a distance of 85 cm is reported in a trial where the robot was started on the outskirts of the gas plume. In such a case, the odour compass pointed to the plume centre first and indicated the direction towards the gas source after reaching the plume. In a further experiment, the gas distribution was modified by an obstacle situated inside the gas plume. Successful trials are also reported under these conditions when the starting position was both inside the plume (tracing distance ≈ 170 cm) and outside the plume (tracing distance ≈ 125 cm).

2.8   Gas Source Tracing Without a Strong Airflow

Without a strong airflow, the detection limits of the available wind measuring devices (anemometers) are not low enough to measure the typically weak air currents in an indoor environment. With state-of-the-art anemometers based on the cooling of a heated wire [ISNM94], the bending of an artificial whisker [RP02] or the influence on the speed of a small rotating paddle [RKK00], reliable readings can only be obtained for wind speeds in the order of at least 10 cm/s. Moreover, there is evidence that wind slower than 5 cm/s is generally hard to detect [Lom86]. As mentioned above, wind fields with lower velocities are often encountered in industrial or domestic buildings. Thus, search strategies that do not require flow sensors and do not depend on a strong airflow are required.

Modification of the environment by means of adding artificial airflow is usually not a feasible option. In most of the cases this would also not be very helpful – if it was acceptable at all. A typical application for a mobile robot would be, for example, to find a gas source that occurs due to an accident or defect. In order to support gas source localisation with such an electronic watchman, all the possible locations of a gas source need to be determined in advance and a fan would have to be placed at each of these locations. It is usually not feasible to ascertain each possible location where an accident might happen, for example, in the case of a gas pipe which may rupture at any point along its length. Additionally, the inclusion of a fan behind a gas source would be highly undesirable in cases where poisonous gases are involved. This is because the fans would need to be oriented towards the floor space, thus aiding the spreading of the dangerous gas.

One possible solution in an environment with weak but approximately constant air currents would be to perform a chemotactic behaviour based on time-averaged concentration measurements. However, this would require long periods of averaging before each step (typically several minutes [INM+01]), resulting in an unacceptably long duration of gas source tracing.

An alternative solution might be provided by the properties of the instantaneous concentration field. There is evidence that turbulent fluid transport creates patterns of spatially distributed eddies, which contain – despite their unpredictable and rapidly changing structure – "several directional parameters for potential use by animals and robots" [CAGC94]. Due to the absence of a theoretical description of turbulence that is able to predict the fine structure of gas distribution in a real world environment, the task is to identify regularities in turbulent distributions (both spatial and temporal) and to figure out how to extract directional and positional information with respect to the source location.

Using Artificial Neural Networks to Predict the Direction to a Gas Source

Machine learning techniques offer a promising possibility for extracting directional cues from the received sensor signal without the need for a theoretically established model of the turbulent dynamics. In addition, a learned system model can also compensate for the lack of an established model to describe the sensor-environment interaction analytically.

Duckett and co-workers used artificial neural networks (see Section 8.6.1) to estimate the direction to a stationary gas source from a series of sensor readings [DAS01]. A beaker filled with acetone was used as the gas source, and the experiments were carried out in an indoor office environment without a carefully controlled airflow (only weak air conditioning).

In a first experiment, training data were acquired by rotating a mobile robot at a distance of 50 cm from the gas source. The robot was equipped with eight metal oxide gas sensors inside a single tube, which was longitudinally divided into two chambers by a solid separation. Two identical sets of four sensors were placed in the two "nostrils", and a small fan was used to pump air through the tube (Örebro Mark I mobile nose). The distance between the two sets of sensors was only a few centimetres. Gas sensor measurements were sampled during a full turn at $10°$ intervals, starting with the air inlet of the sensor tube oriented toward the opposite direction to the source. At each sampling angle, the robot stopped for five seconds. The mapping from normalised sensor readings to the direction of the gas source was then learned with two different neural network architectures using the data obtained in a total of 60 complete turns (50 turns were used for training and 10 turns for testing) and backpropagation.

As a first result, the mean squared error on the test data was found to be higher in the case of a non-recurrent neural network. A validation experiment was then conducted using a recurrent network with ten hidden neurons, eight inputs from the sensors and ten recurrent input units. Here, the source location was estimated from the data recorded in two complete turns. The experiment was repeated over seven runs with different distances to the source. In addition, another seven runs were conducted at the training distance of 50 cm where the robot started to turn facing toward the gas source. While a mean error of about 20 degrees and a standard deviation of 10 to 15 degrees was found in the trials at the training distance and at a smaller distance of 25 cm, the error was rather large at a distance of 75 cm (52.9 ± 58.3 degrees).

In an extension of this work, Farah and Duckett [FD02] implemented a gas source tracing behaviour based on the direction estimates from a neural network. A different design of the mobile nose was used in this experiment because a larger separation between the sensor chambers was found to be advantageous for locating the source. The two sets of metal oxide sensors were mounted inside two separate tubes on either side of the robot, and no fans were used (Örebro Mark II mobile nose). Instead of a recurrent network, a feedforward architecture with one hidden layer of four units was used, which was found to produce the best direction estimates in a preliminary test experiment. The neural network was trained and tested in a circular area with an approximate diameter of 3.5 m, which was constructed from artificial walls in the middle of a laboratory room with moderate air conditioning. Data were collected with the robot driving toward the gas source at a constant speed in seven different directions. The gas source, a beaker with an opening of 7 cm in diameter, was filled with ethanol and placed at four different positions near the border of the circular area. The robot always started in the middle of the circular area at a distance of 80 cm from the source.

Validation experiments, where the rotational speed of the robot was controlled by the output of the feedforward neural network, were carried out using different starting angles and distances. Each trial was stopped when an obstacle was detected. In most of the trials the distance to the gas source was below 50 cm at the end of the trials, while the source was hit exactly in approximately only one third of the trials. Ten trials were performed under the same conditions as during the training phase. An additional ten trials were carried out where the gas source was placed at different positions ("partly-familiar environmental conditions"), the artificial walls were removed ("unfamiliar environmental conditions"), and a different room with no artificial walls was used ("completely new environment"). A significant variation of the tracing performance, however, could not be observed.

The experiments of Farah and Duckett demonstrated the feasibility of the approach to use machine learning techniques in order to predict the direction to the gas source. This prediction was used to reactively steer a robot toward a stationary source. The obtained gas source tracing performance was found to be superior compared to driving in a straight line (i.e., without using gas sensors) with high statistical significance. It would be further instructive to evaluate the performance in comparison with a reactive chemotactic behaviour, i.e., to figure out to what degree the neural net learned to apply bilateral comparison as the optimal solution for gas source tracing in a turbulent environment with the given set-up.

2.9 Comparison of Gas Source Tracing Strategies

Several suggestions for gas source tracing were reviewed in the last two sections. In addition, the gas source tracing strategies that were experimentally tested on a mobile robot in an environment with a strong artificial airflow are summarised in Table 2.1. Apart from the required sensor equipment, a brief description of each implementation is given. The experimental tests and the results obtained are roughly characterised by means of an indication of the wind flow in the experiment and the achieved tracing distance (distance between the start and the end point of successful tracing trials).

It is difficult to compare the suggested gas source tracing strategies due to the variety of the environmental conditions considered [KR03]. Moreover, the software controller cannot be studied independently from the hardware design [FD02] and the sensing strategy [LZWW01b]. For each strategy, successful systems are reported that were able to trace the gas source in a similar time frame. A thorough analysis by statistical means, however, is often not given. Apart from an accurate description of the environmental conditions, a statistical evaluation of the results is especially important for experiments on gas source localisation due to the random nature of turbulent flow. The evaluation of gas source localisation strategies should be based on long-term experiments, preferably containing a comparison of the achieved performance with respect to different environments.

2.10 Approaches Based on an Analytical Model

Regarding the needs of mobile robotics, it is not feasible to model the turbulent wind and gas distribution in a realistic environment with the currently available

Strategy	Min. Requirements / Implementation / Experiment
E. coli algorithm [RBHSW03]	1 gas sensor
	straight "runs" and direction randomising "tumbles", longer runs if concentration increases, higher tumbling frequency otherwise
	uniform air flow ($\approx$ 1.5 m/s), no successful tracing trials
iterative chemo-tropotaxis [RBHSW03]	2 gas sensors
	turn towards the higher concentration by a small angle, move forward, repeat
	uniform air flow ($\approx$ 1.5 m/s), tracing distance $\approx$ 2 m
zigzag approach [ISNM94]	1 gas sensor, directional wind sensor
	upwind plume crossing with turns at the edge of the plume + backtracking
Fig. 2.7 (b)	uniform air flow ($\approx$ 12 or 20 cm/s), tracing distance $\approx$ 80 cm
dung beetle algorithm [RBHSW03]	1 gas sensor, directional wind sensor
	upwind plume crossing with turns at the edge of the plume
Fig. 2.9 (a)	uniform air flow ($\approx$ 1.5 m/s), tracing distance $\approx$ 2 m
plume-centred upwind search [RTDMS95]	1 gas sensor, directional wind sensor
	upwind movement with corrections of the heading according to the concentration gradient across the upwind direction
Fig. 2.8 (a)	uniform air flow ($\approx$ 30 cm/s), tracing distance $\approx$ 1 m
step-by-step progress [ISNM94]	2 gas sensors, directional wind sensor
	simultaneous concentration gradient following across the wind and upwind movement + waiting period (if concentration is low)
Fig. 2.7 (a)	uniform air flow ($\approx$ 12 or 20 cm/s), tracing distance $\approx$ 80 cm
spiral surge algorithm [HMG02]	1 gas sensor, directional wind sensor
	surge in upwind direction with a subsequent local search (outward spiral) triggered by high concentrations
Fig. 2.9 (b)	uniform air flow ($\approx$ 1 m/s), tracing distance $\approx$ 6 m
Bombyx mori algorithm [RBHSW03]	2 gas sensors, directional wind sensor
	local search biased by the wind direction and the concentration gradient, triggered by high concentrations
Fig. 2.8 (b)	uniform air flow ($\approx$ 1.5 m/s), tracing distance $\approx$ 2 m
multiphase tracing algorithm [IKNM96]	2 gas sensors, directional wind sensor
	gradient-following if concentration is low, upwind movement and gradient-following across the wind otherwise + backtracking
Fig. 2.10	non-uniform air flow ($\approx$ 10 – 30 cm/s), tracing distance $\approx$ 1.5 m

Table 2.1: *Gas source tracing strategies that were experimentally tested on a mobile robot in an environment with a strong artificial airflow. The first column shows the identifier of the algorithm as specified by the authors and a reference to a corresponding publication. The second column itemises the required sensors, details of the implementation and the approximate tracing distance (distance between the gas source and the point where the robot was started).*

technology. It is generally a problem that many boundary conditions are unknown. And even with a sufficiently accurate knowledge about the state of the environment it would be very time-consuming to achieve the required degree of accuracy with a conventional finite element model [KR03]. For specific situations, however, the time-averaged distribution can be described in a computationally inexpensive way.

Approaches that utilise an analytical model of the gas distribution provide several benefits, especially the possibility to estimate the properties of the full distribution by fitting local measurements. If the model contains the source position as a parameter it is possible to locate a gas source without moving to it. In contrast to techniques that require following of a gas plume to its source, there is also no need for an additional mechanism to terminate the search in the proximity of the source. On the other hand, model-based approaches can of course only be successful within the scope of the model. It is especially arguable to what degree these approaches are robust against deviations from an average stationary distribution due to turbulence.

Modelling the Gas Distribution in a Unidirectional Constant Airflow

A gas source localisation method that uses a model of the gas distribution was developed by Ishida et al. [INM98]. The method assumes a unidirectional wind field with a constant average wind speed and that turbulence is isotropic and homogeneous. Under these conditions, the time-averaged concentration profile of a point source depends only on the distance from the source, the wind vector, the turbulent diffusion coefficient and the emitting rate of the source (see Eq. 8.1). With the exception of the wind speed, which was measured with the anemometric sensor, the parameters are estimated from the recorded sensor signal using standard nonlinear optimisation methods.

Experiments were carried out in the same clean room where the multiphase algorithm was also tested (see "Multiphase Tracing Algorithm" in Section 2.7). The wind speed was approximately 20 cm/s. First, the sensor response was recorded over 300 s along a z-shaped track that covered an area of approximately 40×25 cm^2 inside the gas plume. By fitting the average signal from the four sensors of the probe shown in Fig. 2.6, it was possible to estimate the location of an ethanol source (75 ml/min) over a distance of approximately 1 m with an error of only 6.4 cm.

A combination of gas source tracing with the model-based estimation of the gas source localisation was implemented in a second experiment. The robot was driven on a zigzag track towards the predicted source position and the estimate was continuously updated from the signal obtained along the driven path. In the reported trial, the gas source was successfully traced over a distance of approximately 1 m. The estimation error fell below 25 cm at a distance of approximately 50 cm from the source and fell to near zero during the approach. However, convergence of the predicted source location to the real one could not be achieved in front of the gas source, which is interpreted by the authors as an effect of three-dimensional airflow that cannot be described by the two-dimensional model.

Modelling the Wind Field Using Naive Physics

An approach that utilises a rule-based model has been suggested by Kowadlo and Russell [KR03]. The central element of the method is a map of the airflow in the environment of the robot, which is created by using a set of naive rules derived from common sense and physical intuition. This approach – using a set of plausible rules instead of classical physics – has been proposed as "naive physics" by the AI community [Hay78].

On the basis of flow characteristics observed in a simulation of laminar air flow in a rectangular room with up to two objects (resembling the actual experimental environment), Kowadlo and Russell developed a set of rules to model the flow field for similar scenarios. Using this set of rules, the authors implemented a reasoning algorithm that uses local measurements and a two-dimensional airflow gridmap in order to determine probable gas source locations. Given an *a priori* map of the room, an airflow map is first generated and possible plumes of gas source candidates (objects in the room) are projected downwind. This initial step creates a list of target positions (one on each possible plume) that are subsequently investigated in order to identify whether the corresponding object is a gas source.

The reasoning algorithm was implemented on the two-wheeled robot Roma, which has a circular shape (diameter: 24 cm) and was equipped with a metal oxide sensor. Ten trials were performed in a "pseudo two-dimensional" room with a small height of ≈ 25 cm compared to the dimensions of the floor area (2.8×1.9 m^2). As potential gas sources, two objects with a height equal to the ceiling and a footprint of 26×26 cm^2 were placed in the room. A laminar airflow of ≈ 50 cm/s was introduced into the room and the gas source was created by injecting ethanol vapour from one side of one box at a rate of 30 ml/min.

Without travelling all of the way to source candidates, the correct object could be identified as being the gas source in all of the ten trials. It remains arguable, however, whether this approach can be extended to a less artificial three-dimensional environment with a turbulent airflow. A further limitation of the approach is the need for an accurate three-dimensional map of the environment. Moreover, only the geometrical properties of the environment are modelled, thus neglecting the influence of temperature and pressure gradients. Note that the proposed algorithm essentially addresses the problem of gas source declaration because the positions of gas source candidates are included in the *a priori* map of the environment.

2.11 Suggestions for Gas Source Declaration

In order to provide a solution to the full gas source localisation task, the gas source tracing strategies reviewed in Section 2.7 and 2.8 need to be extended by an additional mechanism to terminate the searching behaviour. Generally, this means that a gas source localisation method requires some means of determining proximity to a source. By contrast to pure gas tracing strategies, a separate declaration step is not necessary in the case of the model-based approaches introduced in Section 2.10, or the methods of gas source localisation by gas distribution mapping (Section 5.9) and gas source localisation by exploration and concentration peak avoidance (Section 6.5.2) suggested in this thesis. In order to verify the estimated source location, an independent method to determine the certainty that the source has been found would nevertheless be beneficial to any gas source localisation strategy. However, though some suggestions for solving the gas source declaration task have been made in the literature, a detailed analysis based on experimental tests is not yet available, apart from the work presented in this thesis (see Chapter 8).

Gas Source Declaration Using Other Sensor Modalities

Probably the simplest declaration strategy is to assume that a gas source has been found when the robot detects an obstacle during gas source tracing. A source would thus be identified by using a sensor modality different from the gas sensors (bump sensors or a range scanner, for example). Apart from the apparent possibility of misclassifications, a further drawback of this method is that it assumes that the gas source appears as an obstacle to the robot. By using previous knowledge and different sensor modalities, a gas source can also be identified by means of other properties such as its temperature (in the case of a fire) or visual features such as brightness (a fire), shape (a pipe) or a mirror-like reflection (a puddle), for example.

Gas Source Declaration Based on Average Concentration Measurements

If detailed information about the air flow and the intensity of the gas source is given, it should be further possible to estimate the distance to the source from time-averaged concentration measurements. Sufficiently accurate previous knowledge about the properties of an expected gas source, however, will not always be available in advance. It would rather be desirable to exploit more general features for gas source declaration.

Detection of a Concentration Drop at the Source Location

In an environment with a sufficiently strong unidirectional air flow, it is expected that the concentration should be very high on the downwind side of the source but low on the upwind side. This difference on opposite sides of a source can be detected without explicit knowledge of the airflow direction by circumnavigating the source as suggested by Russell and co-workers [RTDMS95]. The method requires that the gas source takes the form of an obstacle. An alternative method in cases where the source cannot be sensed as an obstacle would be to approximate the gas source location by the point where the robot loses the gas plume during upwind search [RTDMS95]. A similar method was suggested by Hayes et al. [HMG02] as an extension of the

Spiral Surge Algorithm discussed in Section 2.7. At the head of a plume, the Spiral Surge Algorithm tends to surge into an area of low concentration, and then spiral back to the origin of the surge before receiving another "odour hit". The vicinity to a gas source is thus expected to appear as a series of small distances between the locations where the robot senses consecutive "odour hits". A further method based on the same characteristic was suggested by Ishida et al. [IM03; IYK+00]. In order to detect the transition from very high to low intensities at the source location, the authors suggest to use a visualisation of a gas distribution in the form of a gray-scale image obtained with a portable homogeneous array of gas sensors. A common disadvantage of these approaches is that a drop in the sensor readings might occur also at locations distant from the source due to the turbulence of the airflow. Furthermore, they rely on a sufficiently strong and constant airflow that gives a low probability for patches of gas to occur on the upwind side of the source.

Detection of Reducing Plume Width

Two of the tracing strategies introduced in Section 2.7 include repeated crossings of the gas plume, changing the direction upon detection of the outer edge of the plume ("zigzag approach method" and "dung beetle algorithm"). The length of the path between subsequent turns provides information about the plume width. It should be possible to estimate the distance to the gas source from the reducing width obtained during the movement in upwind direction [RBHSW03]. This approach rests upon the assumption of a unidirectional air flow. While the two mentioned tracing strategies assume further on that the air flow is strong enough to be detectable with an anemometer, this condition is not essential. In theory, the upwind direction could also be determined without an anemometer by detecting the direction of reducing plume width. However, the practical use of this gas source declaration method, especially regarding the question to what extent reliable information about the plume width can be extracted from the turbulent gas distribution, remains to be validated in a real experiment.

Fine Structure of the Gas Distribution

Further directional and positional clues with respect to a gas source could be provided by the fine structure of the concentration field. Atema reports on underwater measurements of odour dispersal patterns generated from a jet nozzle, which indicate possible features of the turbulent distribution that exhibit such clues [Ate96; CAGC94]. The results indicate that bilateral comparison of correlated peak arrivals could possibly be used to determine the direction of the concentration gradient. In the experiments with a stereo nose architecture, a tendency was found that the gas sensor, located closer to the plume centre (where the concentration is higher on average) was hit first by gas patches that were sensed with both sensors. As a more reliable method to determine the direction of the spatial concentration gradient, the author mentions the onset slope of individual concentration peaks. Closer proximity to the source was found to correspond to a steeper slope. Each gas patch is sheared and diffused while it drifts away from a gas source. Therefore it seems possible that the spatial gradient of peak onset slopes provides a sufficiently general feature of gas dispersal that allows estimation of the distance to a gas source and thus enables a method for gas source declaration. A further possibility could be to distinguish different areas in a plume by the characteristic distribution of peak shapes. However, while there is some evidence that it could be possible to derive directional and positional clues from the fine structure of a turbulent gas plume, the feasibility of such an approach could not be demonstrated experimentally so far.

Chapter 3

Design of the Gas-Sensitive Robots

"There is a real danger (in fact, a near certainty) that programs which work well on simulated robots will completely fail on real robots because of the differences in real world sensing and actuation – it is very hard to simulate the actual dynamics of the real world." (Rodney Brooks [Bro92])

This chapter introduces the mobile robots and the two different gas sensing systems that were used for the experiments discussed in this thesis. First, the gas sensors utilised in both setups are described in Section 3.1. Then, a small scale platform is detailed in Section 3.2, consisting of a Koala mobile robot (Section 3.2.1) and an inexpensive self-made gas-sensitive system, the Örebro Mark III Mobile Nose (Section 3.2.2). In contrast, a medium scale platform is described in Section 3.4. It consists of the ATRV-Jr robot "Arthur" (see Section 3.4.1) and the commercially available gas sensing system VOCmeter-Vario (Section 3.4.2). The discrimination of a "small" from a "medium" scale platform used here does not only apply to the physical dimensions of the robot but also to the complexity (and thus the total cost) of the robot and the gas-sensitive system. The small scale system can be thought of as a study of a commonly available mobile robot that is equipped or can be additionally supplied at low cost with a device similar to the Mark III mobile nose. By contrast, the mobile robot "Arthur" with the considerably more expensive VOCmeter-based gas sensitive system can be regarded as a prototype of a dedicated system that is intended to be used in larger environments or under difficult conditions (like, for example, in a rescue mission). Apart from the size of the robots, however, the most important difference with respect to the experiments carried out in this thesis, is the sensor equipment of the two platforms. Due to the improved self-localisation capability of the "medium scale" robot, it was possible to perform experiments in a larger environment. Although a stronger interaction between the larger robot and the gas distribution is expected, only minor differences of the results obtained with the two platforms were observed in the indoor experiments carried out in this thesis.

This chapter contains a general description of the robots and the design of the mobile noses used. Due to its flexibility, however, especially the gas sensing system on the robot "Arthur" was utilised in different configurations, which are reported in the relevant chapters.

Aside from the description of the hardware used, this chapter also introduces a method to determine the response characteristics of a mobile nose. Considering the Mark III mobile nose as an example, this method is explained in Section 3.3 and indications of a suitable design for a mobile nose, which result from the analysis, are discussed. The characterisation method is composed of a simple experiment (Section 3.3.1) and the corresponding analysis method (Section 3.3.2). The introduced method models the gas-sensitive system as a first-order sensor (Section 3.3.1), which was found to be a good approximation for particular configurations of the Mark III mobile nose. For these configurations it is possible to describe the response by the time constants of exponential rise and decay, which can be roughly estimated without an extensive experimental setup with the described characterisation method.

3.1 Gas Sensors

As a common feature, both mobile noses introduced in this chapter use metal oxide gas sensors. The observed physical effect is the resistance of the surface layer measured at an operating temperature of approximately 300°C, which changes with respect to the concentration of a deoxidising gas. This type of sensor is discussed in-depth in Section 2.1.

The selectivity of metal oxide sensors can be modified only slightly by using different semiconductors or doping with different catalytic metal additives. So, although the sensitivity of the Figaro sensors [Fig] used in this work is optimised for a particular target gas, as indicated in Table 3.1, they all respond strongly to alcoholic substances, which are used as the analyte in the experiments. Even though the signal obtained from the TGS 2610 sensors, which are especially designed to have low sensitivity to alcohol vapours, was lower than for the other sensors, it was sufficiently high for the experiments presented in this thesis. Thus, an array of these sensors is not particularly well suited to discriminate different odours. In the experiments in this work, multiple sensors were therefore only used to obtain a spatial gradient or to increase the robustness of the system.

Model	Typical substances detected	Typical detection range
TGS 2600	hydrogen, carbon monoxide	$1 - 10$ ppm
TGS 2610	propane, butane	$500 - 10000$ ppm
TGS 2620	organic solvents, carbon monoxide	$50 - 5000$ ppm

Table 3.1: *Figaro gas sensors used in this work and their selectivity characteristics.*

Figure 3.1: *The Örebro Mark III mobile nose on a Koala robot in front of a gas source. Two sets of 3 metal oxide gas sensors (one set is visible) were mounted inside the two suction tubes at the rear of the robot.*

Despite their low selectivity, metal oxide sensors were chosen because they are inexpensive, highly sensitive and relatively unaffected by changing environmental conditions like room temperature or humidity.

3.2 Koala Robot and the Mark III Mobile Nose

In order to investigate a minimalistic system, a mobile nose was developed for a K-Team [KTe] Koala robot according to the following design decisions. First, it was decided to obtain the readings from the metal oxide sensors using a direct measurement of resistance, in order to minimise costs. Second, due to the non-directional nature of single concentration measurements, a stereo nose architecture with two tubes or "nostrils" was used (see Figs. 3.1 and 3.2) to measure the spatial gradient of a gas concentration. This is the simplest architecture that can be used to measure the *instantaneous* gradient of a gas, i.e., without requiring path integration, as with a single sensor. Third, to reduce the latency of the sensors, suction fans were mounted inside the tubes. It could be shown in initial experiments that this design reduces the decay time of the system (see Section 3.3). Fourth, these experiments also showed that a further increase in performance could be obtained by separating the two tubes with a "septum" or dividing wall to reduce interference between the

(a)

(b)

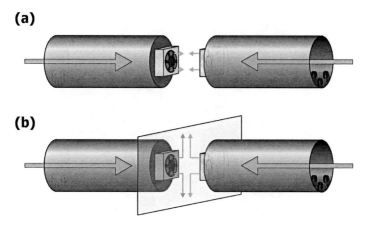

Figure 3.2: *Schematic view of the Mark III mobile nose. The airstream produced by the two suction fans is indicated by arrows (a) without and (b) with a septum.*

opposing airflows. Before these experiments are discussed, the configuration of the Koala robot and the setup of Mark III mobile nose are outlined in the two following sections.

3.2.1 Koala Robot

The six-wheeled Koala mobile robot is shown in Fig. 3.1. It is equipped with 16 infrared proximity sensors. With a footprint of approximately 32×32 cm^2 and a height of approximately 20 cm, it is comparatively small, which implies a minor interference with a gas distribution, especially if the driving speed is kept low. Control programs were executed on a distant host computer which is connected to the robot with a wireless serial link (Bluetooth, RS232). The battery capacity of 4 Ah allows for a total operation time of up to 4 hours. In order to determine the position of the robot, the vision-based absolute positioning system W-CAPS was used, which tracks a distinctly coloured object mounted on top of the robot (the {green} cardboard "hat" shown in Fig. 3.1). The positioning system uses four web-cameras mounted at fixed positions to triangulate the (x,y) position of the centre of the cardboard "hat". By combining up to 6 single position estimates, it provides centimeter level accuracy. Using additional information from the odometry estimates of the robot, an even higher level of accuracy could be achieved in the experiments. The absolute positioning system W-CAPS is detailed in Section B.

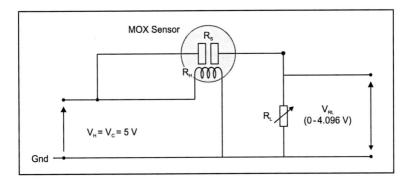

Figure 3.3: *Electrical circuit used in the Mark III mobile nose to measure the resistance of the metal oxide gas sensors.*

3.2.2 Set-up of the Örebro Mark III Mobile Nose

The Mark III mobile nose[1] consists of six metal oxide gas sensors, which were placed symmetrically in sets of three inside two separate tubes containing a suction fan each (see Fig. 3.2). The number of sensors is limited by the number of analog inputs available on the Koala robot. On both sides, metal oxide sensors of type TGS 2600, TGS 2610 and TGS 2620 were used. The sensor arrays were mounted at the outer end of the tubes in order to maximise the instantaneously measurable spatial gradient (see Figs. 3.1 and 3.2). The distance between the two sets of sensors was 40 cm.

To obtain the signal output, the electrical circuit shown in Fig. 3.3 was used. A load resistor R_L is connected in series with each gas sensor and the voltage across the load resistor V_{RL} is measured. This value is then converted by a 10 bit A/D converter that covers an input range of 0 – 4.096 V, and transferred to the host computer using the available wireless serial connection. The voltage across the load resistor depends on the resistance R_S of the surface layer as

$$V_{RL} = \frac{R_L}{R_L + R_S} V_C. \qquad (3.1)$$

V_{RL} depends linearly on the conductance ($\propto R_S^{-1}$) if $R_L \ll R_S$. However, the dependency betweeen the inverse of the resistance of the surface layer and the concentration of an analyte gas is generally not linear. A typical response profile of the output voltage V_{RL} versus gas concentration is shown in Fig. 3.4. In order to obtain

---

[1]The Mark III mobile nose is an extension of the Mark II mobile nose introduced in [FD02].

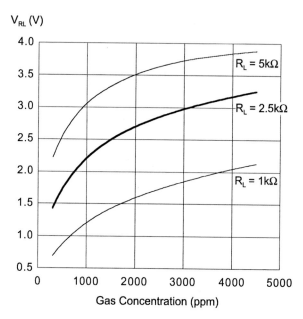

Figure 3.4: *Typical profile of output voltage V_{RL} versus gas concentration in a metal oxide sensor for three different load resistor values R_L. (The graph is taken from the Figaro brochure "General information for TGS sensors" [Fig] where the exact type of gas sensor and the used analyte gas is not specified.)*

an optimal resolution and to be able to modify the mapped range according to the intended application, an adjustable load resistor R_L (1 – 100 kΩ) was used.

The operating temperature is attained and perpetuated by the heat dissipation in the heater circuit (R_H: resistance in the heater circuit), which is powered with the same voltage as the measurement circuit ($V_H = V_C$). Both the voltage supply for the heater and the measurement circuit ($V_H = V_C$) are provided by an additional battery on the robot. Before each experiment only this battery was switched on in order to heat the sensors for at least 45 minutes until they reach their working temperature.

In order to increase the spatial differentiation of the mobile nose, suction fans (Papst 405F [Pap]) that generate an airflow of 8 m³/h were mounted inside the tubes. Thus, the exchange rate of air at the sensor's location is increased and possible effects due to different driving speeds are reduced. As detailed in Section 3.3 below, the decay time could also be decreased with the addition of fans.

3.3 Characterisation of a Mobile Nose

This section introduces a method to characterise the response of a mobile nose *as a whole* as a first-order sensor. It consists of a simple experiment, which can be carried out without too much effort, and an analysis method that is able to compensate for inaccuracies in the realisation of the experiment. This technique is discussed here considering the Mark III mobile nose as an example, but can be applied to any mobile nose.

3.3.1 Characterisation Experiment and Sensor Model

In order to determine the dynamic response of the Mark III mobile nose, the following experiment was performed: alternately one set of sensors was exposed to a step stimulus, which was approximated by opening a bottle of ethanol in the direct vicinity of the sensors for a fixed period of time. For each of the 4 possible configurations that result from using or not using the fans and separating or not separating the two tubes with a septum, the following steps were repeated:

- wait for 20 s,

- open a bottle of ethanol for 10 s at a distance of approximately 1 cm alternately in front of the left- and right-side sensor set,

- close the bottle and wait for another 120 s.

The readings were analysed by fitting a sensor model to the values recorded during each trial. An example of a fitted model together with the corresponding gas sensor readings is shown in Fig. 3.5. Due to the complexity of the interaction between metal oxide sensors and their environment, no physically justified general description of this process is available. It is, however, sufficient for our concerns to assume an ideal first-order sensor and thus model the dynamic response to a step stimulus as an exponential rise and decay. To this end, the applied model (see Eqn. 3.2 – 3.6) separates into three parts according to the three regions shown in Fig. 3.5.

$$R(t) = \begin{cases} R_I(t) & \text{if } t < t_S \\ R_{II}(t) & \text{if } t_S < t < t_S + \Delta t \\ R_{III}(t) & \text{if } t_S + \Delta t < t \end{cases} \tag{3.2}$$

$$R_I(t) = R_0 \tag{3.3}$$

$$R_{II}(t) = R_0 + (R_{max} - R_0)(1 - exp(\tfrac{-(t-t_S)}{\tau_r})) \tag{3.4}$$

$$R_{III}(t) = R_0' + (R_{max}^* - R_0')exp(\tfrac{-(t-t_S-\Delta t)}{\tau_d}) \tag{3.5}$$

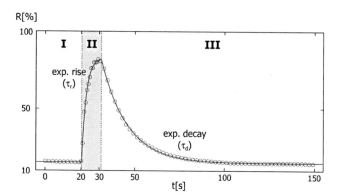

Figure 3.5: *Real gas sensor readings (circles) together with the corresponding fitted dynamic response model (line). Relative output values are shown, recorded in an experiment where fans were used. Notice the three labelled regions of the applied model function (Eqn. 3.2 – 3.6). Note also that 7 out of 8 readings were omitted in regions I and III, and 2 out of 3 in region II for clarity of the plot.*

$$R_{max}^* = R_0 + (R_{max} - R_0)(1 - exp(\frac{-\Delta t}{\tau_r})) \qquad (3.6)$$

A first-order sensor model was found to be a very good approximation if the fans were switched on. This can be seen in Fig. 3.5. With activated fans, the characteristics of the mobile nose can thus be described fairly well using only the time constants of rise τ_r and decay τ_d.

3.3.2   Data Analysis

The sensor model contains a total of seven adjustable parameters. Apart from the time constants of rise and decay, the remaining parameters refer to the actual realisation of the characterisation experiment: the response level R_0 before and after the stimulus R_0', the saturation level R_{max}, the time the sensor started to respond t_S, and the duration of the rising period Δt. To determine all parameters, the model was fitted to the approximately 600 data points recorded per trial using the Levenberg-Marquardt method [Mar63]. This method is an iterative nonlinear least-squares minimisation algorithm. It provides the capability to continuously switch between a steepest descent method, which is used far from the minimum, and the inverse-Hessian method used near to the minimum [PFTV92].

Starting from a reasonable initial guess, the fitting process for each trial leads to a set of parameters together with the corresponding asymptotic standard error.

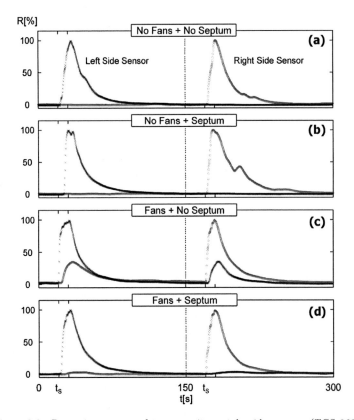

Figure 3.6: *Dynamic response of two opposite metal oxide sensors (TGS 2600) to two consecutive step stimuli on alternate sides of the robot.*

Due to the variations in the experimental conditions, an overall estimate cannot be obtained in a straightforward way by fitting to all of the recorded runs simultaneously. Thus, the overall estimate has to be computed from single results. This is of course only meaningful for parameters that do not change strongly with varying experimental conditions. Assuming the fit parameters for each trial to be drawn from normal distributions with the same mean but a possibly different standard deviation, the maximum likelihood estimate equals a weighted average with the weights being the quadratic inverse asymptotic standard error of the corresponding fit parameter. Thus, for a particular set of trials, the overall estimate of a fit parameter

$\hat{F}$ and the standard deviation $\hat{\sigma}_F$ is computed from the single fit results $F^{(i)}$ and the corresponding asymptotic standard deviation $\sigma_F^{(i)}$ as:

$$\hat{F} = \frac{\sum_i w_i F^{(i)}}{\sum_i w_i}, \tag{3.7}$$

$$\hat{\sigma}_F^2 = \frac{\sum_i w_i}{(\sum_i w_i)^2 - \sum_i w_i^2} \sum_i w_i (F^{(i)} - F)^2, \tag{3.8}$$

$$w_i = \left(\frac{1}{\sigma_F^{(i)}}\right)^2. \tag{3.9}$$

The results for the six sensors used are summarised in Table 3.2, including the maximum likelihood estimate of the duration of the rising period as well as the time constants of rise and decay. Each of these values was calculated as described above from the fit parameters for those trials where the stimulus was placed on the same side as the corresponding sensor. Twelve such trials (six on each side) were performed for each configuration except for the one with fans and septum, which was mainly used for the experiments in this work. With this configuration a total of 18 trials was performed. A characteristic response to two consecutive step stimuli is shown in Fig. 3.6 for each configuration tested.

The last column in Table 3.2 contains the difference between the response time t_S of the sensor on the side on which the stimulus was generated, and the response

Config.	Type	$\widehat{\Delta t}$ [s]		$\hat{\tau}_r$ [s]		$\hat{\tau}_d$ [s]		$\lvert\widehat{\Delta t_S}\rvert$ [s]
		Left	Right	Left	Right	Left	Right	
No Fans	2600	11.0±1.9	9.1±1.7	2.3±0.9	1.4±1.4	15.6±1.3	15.2±3.4	-
&	2610	12.6±1.9	7.7±2.8	0.8±0.7	0.3±0.4	19.6±3.0	13.0±2.3	-
No Septum	2620	12.4±2.3	8.2±2.1	1.5±1.5	2.4±0.8	29.2±4.7	28.5±7.3	-
No Fans	2600	8.4±3.1	7.1±3.7	2.0±0.8	1.0±0.5	15.3±2.2	20.7±5.0	-
&	2610	8.8±3.1	7.5±1.6	2.8±1.3	0.6±0.3	16.6±5.5	18.2±5.5	-
Septum	2620	14.2±14.2	23.4±13.7	4.1±1.3	2.1±0.6	19.8±6.4	36.7±4.6	-
Fans	2600	11.2±1.2	10.5±0.6	2.1±0.8	1.9±0.8	12.7±1.1	15.0±1.4	3.4±0.8
&	2610	10.2±0.7	9.9±0.5	2.5±1.1	1.1±1.1	9.2±1.3	9.6±2.8	3.2±0.4
No Septum	2620	10.4±0.6	10.6±0.5	2.3±1.0	1.4±0.4	7.8±0.9	12.5±0.5	3.9±1.2
Fans	2600	9.8±1.8	10.4±0.8	0.3±0.4	2.3±0.4	12.9±1.3	13.6±0.6	6.3±2.9
&	2610	9.7±1.1	9.5±1.0	2.2±0.9	1.3±0.6	8.8±0.9	10.9±3.1	5.5±3.2
Septum	2620	7.9±3.6	10.0±0.6	2.3±1.2	1.5±0.6	9.3±2.7	10.5±1.5	6.0±4.0

Table 3.2: *Summarised fitting results of the dynamic sensor response to a step stimulus. The maximum likelihood estimate of the rising period $\widehat{\Delta t}$, the time constants of rise $\hat{\tau}_r$ and decay $\hat{\tau}_d$ as well as the delay in responding $\widehat{\Delta t_S}$ is given for each sensor.*

time of the sensor's counterpart on the other side. When the fans remained switched off, the remote sensor's response was too weak to obtain a meaningful fit. Therefore the last column contains values only for those experiments where fans were used.

Analysing the values in Table 3.2 shows that fitting with the assumed model is able to reproduce the stimulus' duration. The agreement with the intended time of 10 seconds is quite good, considering the fact that the actual effective duration depends on varying factors such as the local airflow or how the bottle was opened and closed. The obvious outliers for the TGS 2620 sensors (indicated by a large asymptotic standard error) in the configuration without fans and a septum are caused by the fact that in all the corresponding trials the response declined extremely slowly and unsteadily. Thus, the applied fitting algorithm was not able to ascertain the first clear drop at the end of the rising period II.

Next, it is interesting to compare results of measurements where fans were used with those where the fans remained switched off. Checking the rising time constants $\hat{\tau}_r$ using a paired Student's t-test [PFTV92] shows no statistically significant difference with respect to the usage of fans ($p_{H_0} = 0.9960$). By contrast, the decay time constants $\hat{\tau}_d$ turn out to be clearly higher in the case where no fans were used. The opposite hypothesis can be rejected with high statistical significance ($p_{H_0} = 0.0015$).

A rough estimate of the mobile nose response characteristics is given by the average of the estimated time constants over all sensors.

$$\tau_r \approx 1.8 \text{ s} \tag{3.10}$$

$$\left(\tau_d^{(NoFans)} \approx 20.7 \text{ s} \right) \qquad \tau_d^{(Fans)} \approx 11.1 \text{ s} \tag{3.11}$$

The time constant $\tau_d^{(NoFans)}$ is given in brackets to indicate that the parameter estimates, obtained in the trials where fans were not used, have to be treated with care, because the fitted curve was generally less suitable to approximate the measured values compared to the trials where the fans were switched on. Note further that the rise and decay constants depend generally on the sensor type. In addition, these characteristics vary between different sensors of the same type, and also for one sensor over prolonged periods of time. Finally, they also depend on the gas concentration. Bearing these restrictions in mind, the approximation given in Eqns. 3.10 and 3.11 provides a reasonable notion of the mobile nose's characteristics: the use of fans does not influence the response time to a presented stimulus, but rather lowers the time needed for the sensors to recover after the stimulus has been removed. This is caused by the higher rate of air exchange effected by the fans.

On the other hand, an increased exchange of gas provokes also a less clear distinction of the measured response with respect to the sensors location. Although the air streams caused by the fans are directed against each other (see Fig. 3.2), a considerable amount of gas is transported from one side to the other if no solid separation is used inbetween the suction tubes. This is indicated by the relative strength of the response, as can be seen by comparing Fig. 3.6 (c) and Fig. 3.6 (d),

Figure 3.7: *The gas-sensitive robot "Arthur" in front of a gas source. Inside the robot's body, the base unit of the gas sensing system VOCmeter-Vario can be seen. The figure also shows eight gas sensors (indicated by arrows) connected to the base unit over thin coaxial cables.*

and by the delay of the response time t_S. Again, a statistically significant difference could be assessed by testing on the equality of the calculated time difference $\widehat{\Delta t_S}$ (paired Student's t-test: $p_{H_0} = 0.011$).

Although the values given in Eqns. 3.10 and 3.11 provide only a rough approximation of the reponse of the Mark III mobile nose, the implication of the described experiment is nevertheless apparent: in order to get measurements that reflect the instantaneously sensed concentration as closely as possible, the usage of fans is clearly favourable. However, the airstream produced by the fans must then be separated carefully as it is done, for example, in the Mark III mobile nose by using a septum.

3.4 The Gas-Sensitive Robot "Arthur"

3.4.1 Robot "Arthur"

The mobile robot "Arthur" shown in Fig. 3.7 is based on the outdoor model ATRV-Jr from iRobot [iRo], which is a four wheeled, skid steered vehicle. It is equipped with 17 sonar sensors and a 12 Channel GPS receiver. The sensor equipment was extended by adding the gas-sensitive system described in Section 3.4.2, a SICK laser range scanner [SIC], and a stereo camera[2] mounted on a Pan-Tilt unit (PTU-46-17.5) from Directed Perception [Dir]. For the indoor experiments presented in this thesis only the odometry and the laser scanner were used in addition to the gas sensors. Depending on the number of sensors used, the battery capacity of 66 Ah allows for an operation time of 2 – 3 hours.

With a length of 78 cm, a width of 64 cm, and a height of 55 cm (without laser scanner, Pan-Tilt unit and stereo head[3]), "Arthur" is considerably larger than the Koala robot. An evidently stronger disturbance of the measured gas distribution, however, could not be observed in the experiments. This might be partly a result of the fact that all attempts were made to minimise the influence of the robot on the gas distribution. For example, a device to deflect the air stream that is produced by two rear fans was mounted at the back side of "Arthur". This device can be seen in Fig. 3.7. Apart from that, a low maximum speed of 20 cm/s was applied in all the experiments presented here.

In contrast to the Koala robot, it is possible to execute control programs directly on the robot on a standard PC with an 850 MHz Pentium III double processor. The available wireless LAN connection from ARtem [ARt] (IEEE 802.11b, 11MBit/s) was only used for monitoring and during the development phase.

3.4.2 Gas Sensing System VOCmeter-Vario

As an easy to use off-the-shelf product, the commercial gas sensing system VOCmeter-Vario from Applied Sensor [App] was chosen. It consists of two main components, a small-sized base unit ($19 \times 12 \times 6\ cm^3$) and up to eight gas sensors that are connected to the base unit by thin coaxial cables (RG158). The base unit is shown in Fig. 3.8 (a) and can also be seen inside the robot's body in Fig. 3.7. It collects the readings of the connected sensors at a rate of up to 4 Hz and transfers them to the host computer over an RS-232 serial link. The gas sensors are embedded into so-called sensor sticks, which are tubes with a length of 50 mm and a diameter of 10 mm that contain the actual sensor unit and the required transducing electronics. Two sensor sticks can be seen in Fig. 3.8 (a), while Fig. 3.8 (b) shows a

---

[2]The stereo camera was first composed of a synchronised pair of CCD cameras (XC-999P) from Sony [SON] and then exchanged with the stereo head STH-MD1-C from Videre Design [Vid]. Thus, different camera devices can be seen on the pictures of "Arthur" shown in this thesis.

[3]The total height with laser scanner, Pan-Tilt unit and stereo head is 93 cm

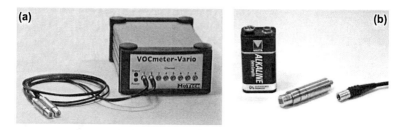

Figure 3.8: *The gas sensing system VOCmeter-Vario. (a) Base unit and two of eight gas sensors (sensor sticks) connected with thin coaxial cables. (b) Disconnected sensor stick together with a standard 9 V battery to demonstrate its size.*

disconnected sensor stick together with a standard 9 V battery to demonstrate its size. This modular design provides a flexible setup that facilitates positioning and exchanging of individual gas sensors and allows quick swapping between different types of sensors. The sensor sticks are available containing a metal oxide (MOX) or a quartz micro balance (QMB) sensor unit. For the considered analytes and concentrations, however, QMB sensors were found to be not sensitive enough.

In contrast to the Mark III mobile nose, fans were not used to ventilate the sensors in the Vocmeter-Vario based gas sensing system. A further difference is the way in which the resistance of the surface layer of a metal oxide sensor is determined. While the resistance is measured in the Mark III mobile nose using a load resistor connected in series with the sensor, a more sophisticated technique is applied in the sensor sticks, which allows access to a considerably larger dynamic range.

3.5   Summary

Apart from a method to determine the response characteristics of a mobile nose, two gas-sensitive mobile robots were introduced in this chapter: a small scale platform (Koala robot and Mark III mobile nose) and a medium scale variant (ATRV-Jr robot "Arthur" and the gas sensing system VOCmeter-Vario). The latter robot was used for the gas source localisation experiments in a "one-dimensional" environment presented in Chapter 4, an investigation of the applicability of a moth gas source localisation strategy for a mobile robot (Chapter 7) and the gas source declaration experiments (Chapter 8). The small scale platform was utilised for concentration mapping (Chapter 5) and the experiments to investigate the performance of Braitenberg-type reactive gas source localisation strategies (Chapter 6).

Chapter 4

Gas Source Localisation in an Environment Without a Strong Unidirectional Airflow

"One of the interesting features of olfaction is the close interaction between olfaction and behaviour." (Hiroshi Ishida [IM03])

A starting point of this thesis was the idea of a mobile robot that is able to perform the tasks of a gas-sensitive electronic watchman – especially the task of gas source localisation – without being restricted to an environment with a strong unidirectional airflow. While several publications are available that address the problem of gas source localisation under the assumption of a constant airflow in the order of at least 30 cm/s (see Chapter 2) the feasibility of gas source localisation under less restricted conditions was unclear at the beginning of this work. Therefore, the initial experiments were designed to investigate fundamental questions concerning the use of gas sensors on a mobile robot in a real world indoor environment without ventilation. These experiments and the corresponding results are summarised in this chapter.

With regard to a typical task for an electronic watchman, the gas source was intended to resemble a small puddle of a leaking liquid chemical. This was achieved by using containers of different size filled with ethanol or acetone. These alcoholic substances were used because they are volatile and non-toxic. Furthermore, they can be regarded as having a typical odour in the sense that liquid chemicals frequently contain fractions of alcohol.

The first question is therefore whether the available sensors are sensitive enough to detect such a typical gas source in a real world scenario. This includes also the question at what distance the presence of a gas source can be perceived. A second main question is then whether the point of minimal distance between the source and the path of a robot corresponds to a maximum in the response of the gas sensors. It appeared that while in some experiments such a correspondence could be observed, the maximum response occured far away from the gas source in other trials. This raises further questions about the possibility to optimise the experimental procedure and the set-up of the gas sensing system in order to achieve a stronger correlation between sensor response and proximity to a source.

For the first experiments, two similar corridor scenarios were chosen. In either case, a self-contained, abandoned environment was considered, meaning that the doors and windows were kept closed and persons were not allowed to enter the room. During each trial the robot was driven up and down the corridor (without turning at the end) while keeping track of its middle. Referring to this movement along a single axis on which the gas source was also located, the scenario can be described as "one dimensional". The corresponding experiments are described and discussed in Section 4.1.

In an extension of the "one dimensional" experiments, Section 4.2 reports on experiments in a "two dimensional" scenario. Here, the correlation between the location of the gas source and response peaks obtained along a path that covers a rectangular area was analysed.

4.1   "One Dimensional" Scenario

The experiments described in this section were carried out in two slightly different corridor environments. A first set of trials was performed in an untenanted apartment without airing ("Corridor–1"). The length of this corridor was 25 m and its width was 2.5 m. During the experiments, the doors and windows were kept shut and no people entered the corridor.

A second series of measurements was performed in a corridor at a university building ("Corridor–2"). While the dimensions of the corridors are comparable (the length of "Corridor–2" was 40 m and its width was 2.2 m), the two "one dimensional scenarios" considered differ slightly concerning the degree of seclusion. "Corridor–2" is weakly ventilated (the air is expelled at the ceiling), one end is completed by a frequently used door and it was not possible to prevent some people walking by (although the experiments were carried out in the night for this reason). The gas distribution in scenario "Corridor–2" was thus more affected by external disturbance.

In each trial, sensor readings were collected along a path, running along the centre of the corridor. A version of the applied control program with turning at the end of the corridor is discussed in Section A.2 and the corresponding DDFLat diagram is shown in Fig. A.5.

In order to emulate gas concentrations comparable to small puddles of leaking liquid chemicals, a container filled with ethanol or acetone was used as a gas source. Various intensities were generated by using three different containers with a cross section of 20, 60 and 130 cm^2 respectively. Approximately one hour before each

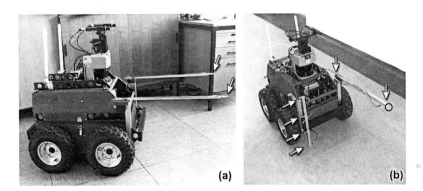

Figure 4.1: *The gas-sensitive robot "Arthur" as it was used in the experiments in a "one dimensional scenario". Sensor positions are marked by arrows and the intake of the tube over which the measuring cell is supplied with gas samples is indicated by a hollow circle. The configuration shown in (b) was also used in the experiments in a "two dimensional environment".*

trial, a container was filled in a separate room and covered with a plate. Then, it was placed either at one end of the driven path or in the middle of it. Because the latter configuration requires the robot to drive over the container, only the smallest container could be used in the corresponding trials. Finally, the experiment was started with a few runs up and down the corridor where the container remained covered (reference measurements). Without interrupting the robot's movement, the gas source was then uncovered at a certain time, which defines the beginning of the actual experiment ($t = 0$ s).

4.1.1 Experimental Set-Up

The experiments were carried out with the gas-sensitive robot "Arthur" that was described in Section 3.4. Two different configurations for the gas sensing system were used for the experiments described in this section as shown in Fig. 4.1. First, the robot was equipped with a pair of aluminium bars with a length of 60 cm (see Fig. 4.1 (a)). Driven by servo motors, these "antennae" could be rotated about a vertical axis near the robot's front. Thus, it was possible to sample the gas concentration at different points in the vicinity of the robot without moving. This feature, however, was not found to be beneficial, because it appeared in the experiments discussed below that the correlation between sensor response and proximity to a gas source is generally less pronounced if the robot is stopped for the measurements. Therefore, the bars were mainly used as a stiff extension, enabling the placement

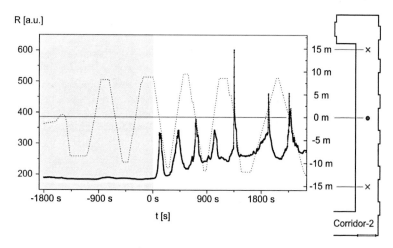

Figure 4.2: *Comparison of gas sensor readings obtained before (t < 0 s) and after the source was uncovered (t > 0 s). Both the sensor readings and the position of the robot are plotted referring to the left and right ordinates, respectively. The period before the container was uncovered is indicated with a light shading.*

of the gas sensors at a distant position in front of the robot, with the intention of minimising the influence of disturbances caused by the robot. Two TGS 2620 metal oxide sensors were either placed at the front end of the bars (as indicated in Fig. 4.1 (a) by the arrows) or directly on the robot's front bumper. Thus, the coordinates of the gas sensors with respect to the projection of the centre of mass to the floor[1] were approximately (80 cm, ±25 cm, 47 cm) on the bars or (40 cm, ±20 cm, 10 cm) at the front bumper.

For the second series of measurents (in "Corridor-2" and also in the "two dimensional" environment) the configuration was modified. Here, eight sensors of the same type TGS 2620 were used. Four of them were mounted on a vertical rod (40 cm, -20 cm, 5/18/33/49 cm) and two on a single bar (35/80 cm, 25 cm, 47 cm). The sensor positions are indicated in Fig. 4.1 (b) by arrows. Two further TGS 2620 sensors were placed inside a cell, which is pumped by a PTFE tube whose intake is located on the outer end of the outstanding bar (82 cm, 25 cm, 47 cm). The intake is indicated by a hollow circle in Fig. 4.1 (b). Due to the small distance to the outer sensor on the bar, it is possible to sample the gas concentration in this range in two

_____

[1]As usual, the coordinate system assumes that the x-axis is oriented with the heading of the robot and the z-axis points upwards perpendicular to the floor.

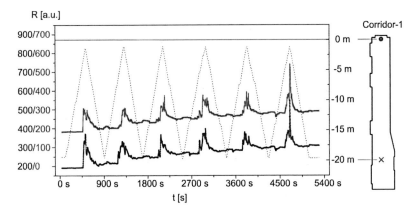

Figure 4.3: *Gas sensor readings recorded while the robot was driven at a constant speed of 5 cm/s. The readings of two sensors are displayed, which were mounted at the front end of the aluminium bars that can be seen in Fig. 4.1 (a). For clarity of the plot, the sensor values, which refer to the left ordinate, are plotted against two shifted scales as indicated by the labels on the left side.*

different ways either with a sensor, which is directly exposed to the environment, or with another sensor operated in a cell with a well-defined airflow.

4.1.2 Results

As a first result, QMB sensors were found to be not sensitive enough to detect the presence of a gas source in the scenario defined above. By contrast, the signal obtained with metal oxide sensors allows us to clearly discriminate the measurements before and after the gas source was uncovered. This can be seen in Fig. 4.2, which shows an experiment that was carried out in "Corridor–2". Both the readings obtained with the undermost gas sensor mounted on the vertical rod and the position of the robot are plotted in the graph against the time since the container was uncovered. The position of the robot is indicated with a broken line referring to the scale on the right ordinate while the sensor readings refer to the scale on the left ordinate. In the experiment, the smallest container (20 cm^2) filled with acetone was used. Despite the rather low analyte concentration observed, it is nevertheless apparent from the plot that the readings obtained before the source was uncovered ($t < 0$ s) differ strongly from those recorded afterwards.

A further observation of the experiments in a "one dimensional scenario" is the minor influence of the actual sensor position with regard to the suitability for gas source localisation. An evident difference in this respect could not be established. A comparison of the response of two sensors mounted at the front end of the aluminium bars (see Fig. 4.1 (a)) is displayed in Fig. 4.3. This experiment was carried out using a container with a cross section of 130 cm² filled with ethanol. The container was placed at one end of the corridor in order to prevent the robot from toppling it. Due to the required obstacle clearance, the minimal distance between the metal oxide sensors and the gas source was approximately 50 cm in this experiment. In the same way as in Fig. 4.2, the sensor readings and the position of the robot (broken line) are plotted referring to the left and right ordinate respectively. For clarity of the plot, the sensor values are plotted against two scales shifted against each other by a constant value. The course of the readings, however, is very similar although the distance between the sensors is approximately 50 cm.

An astonishing outcome of the experiments is that a direct correlation between the maximal sensor signal and proximity to the gas source could be observed only if the sensor readings were recorded while the robot moved at a constant, sufficiently high speed. An example of a trial where the mentioned correlation could be observed is shown in Fig. 4.3. The sensor readings were collected in this experiment while the robot was driven with a constant speed of 5 cm/s. The striking difference compared to the course of sensor readings collected with a stop-sense-go strategy appears in Fig. 4.4. Again, the experiment was carried out in "Corridor–1" using a container with a cross section of 130 cm² filled with ethanol and placed at the end of the corridor. Both parts of the figure display readings obtained in the same experiment with the same sensor mounted at the end of one of the aluminium bars. As in Figs. 4.2 and 4.3, the sensor readings and the position of the robot (broken line) are plotted in the same graph. In the upper part, sensor readings are shown that were recorded while the robot applied a stop-sense-go strategy, moving one metre each step. Consequently, the sensor measurements were mainly obtained while the robot was not moving. By contrast, the readings in the lower part of Fig. 4.4 were collected while the robot was driven with a constant speed. A strong correlation between maximal sensor response and closest proximity to the source appears only in the latter case. The same effect could be observed in several experiments in both corridors and was also confirmed by Farah [FD02]. A weak correlation appears also if a low driving speed of the robot is chosen, similar to the first up-and-down segment shown in the lower part of Fig. 4.4 where the robot was driven at a constant speed of 5 cm/s. Here, the location of the gas source is not indicated as clearly as in the response curve of the sensors, compared to the subsequent trials, which were carried out at speeds of 10 cm/s and 15 cm/s. Note that, except from the last up-and-down movement, the sensor readings shown in Fig. 4.2 were also collected while the robot was driven at a constant speed of 10 cm/s. While the correlation between close proximity to the source and maximal sensor response is not especially evident in the first segment in the lower part of Fig. 4.4, a speed of 5 cm/s was indeed found to be sufficient in most of the trials where this speed was set. This can be seen in the last

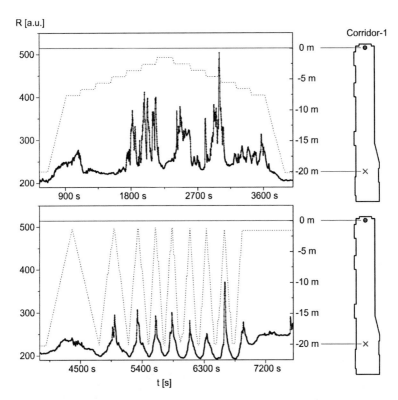

Figure 4.4: *Comparison of gas sensor readings obtained in an experiment where a stop-sense-go strategy was applied first (top) and later on gas sensor readings were collected while the robot was moving with a constant speed (bottom).*

up-and-down movement displayed in Fig. 4.2 and also in the trial shown in Fig. 4.3, which were carried out with this speed. In experiments where the robot moved at a lower constant velocity, however, a strong peak-to-source-location correlation could not be observed. Likewise, such a correlation was also not found in all the trials where a stop-sense-go strategy was applied.

It is not completely clear why a strong correlation between sensor response and proximity to a gas source cannot be achieved if a stop-sense-go strategy is applied. As a possible reason, it was assumed that the analyte material, which is consumed due to the combustion process at the sensor surface, is not replaced fast enough if the robot

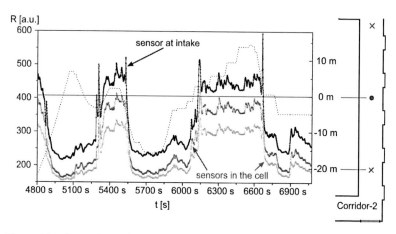

Figure 4.5: *Comparison of gas sensor readings recorded with and without an additional airflow.*

is not moving. The sensor signal would then be dominated by weak airflows that supply the analyte material, rather than by the actual gas concentration. It could be shown, however, that this speculation is not correct. A plentiful supply of the analyte gas is found either if the robot moves at a sufficiently high speed or if an additional airstream is generated at the sensor location. While the robot is not moving, the sensor readings collected inside the pumped cell should thus differ strongly from those obtained with a sensor that is exposed directly to the environment. Such a difference could not be observed. A comparison of measurements collected with two sensors inside the pumped cell and readings obtained with a sensor that is mounted in close vicinity to the intake of the tube, over which the measuring cell is pumped, is shown in Fig. 4.5. During the trial, the cell was pumped at a constant rate of 300 ml/min, while the robot was manually driven and stopped at several points in the vicinity of the gas source. Fig. 4.5 is composed in the same way as Fig. 4.3, except that the sensor readings are slightly shifted against each other referring to the same scale for clarity of the plot. Note that the signal obtained with the sensors in the measuring cell is shifted along the time axis by a constant value in order to compensate for the delay due to the transport in the tube. The course of the sensor response, however, reveals a highly similar trend for all of the sensors. This result is consistent with further tests where a PC fan was mounted in front of a gas sensor to generate an additional airflow at the sensor's surface. Consequently, the consumption of analyte material does not account for the poor suitability of a stop-sense-go strategy for gas source localisation.

Another explanation that has been suggested by Farah [FD02] is that the deceleration of the stopping vehicle itself causes extra airflow, which affects the sensor readings. Because the effect was observed with robots of considerably different size and also while a constant but very low speed was applied, this explanation can be also excluded with high certainty.

A possible explanation for the constant velocity effect is the implicit integration over subsequent measurements, which is performed by metal oxide sensors due to their long decay time (see Sections 2.1 and 3.3). As a consequence, the many local concentration maxima around the gas source appear as a single peak in the readings if the robot is driven at sufficiently high speed. Due to the higher density of local maxima in the vicinity of the gas source, the peak is likely to be centred near the source location. As a consequence of the delayed response of metal oxide sensors, it is further expected that the centre of the response peaks is asymmetrically shifted towards the direction of movement. The response time is, however, substantially lower compared to the decay time and thus a relatively small shift in the order of approximately 2 s times the applied constant velocity should be expected. An asymmetric shape of the observed peaks due to the long decay time is, on the other hand, not critical as long as only the local maximum is used to estimate the location of the gas source.

While the experiments show that it is feasible to localise a gas source in a "one dimensional scenario" by means of a single peak in the sensor response, it has to be emphasised that the sensor peak provides only a comparatively rough approximation of the source location. In the trials where the gas source was placed in the middle of the robot's path (see Fig. 4.2), for example, the average distance to the source at the time of the maximal sensor response was (99.9 ± 159.5) cm, while some distances over 3 metres occured.

The presence of a gas source can be assessed by validating that the sensor readings exceed a certain threshold value. Nevertheless, it is probably more practical to use a surge in sensor response as an indicator for a gas source. Thus, the costly calibration procedure of metal oxide sensors can be avoided and slowly changing sensor characteristics due to ageing of the sensors or varying environmental conditions like temperature and humidity can be compensated. The beginning of the peak, which indicates the gas source location, could be observed up to a distance of approximately 5 metres from the source in the experiments where the robot moved at a constant speed. However, because of the rising level of analyte concentration in a substantially self-contained environment, it is expected that a robot might perceive a surge at a larger distance if it enters a room with a gas source that has been active for some time. The rise of the overall concentration during a trial can be seen in Figs. 4.2 and 4.3.

Finally, the similarity of the results obtained in two different corridor scenarios and with different experimental set-ups has to be emphasised. The presence of the gas source appeared clearly in the profile of the sensor readings regardless of the size

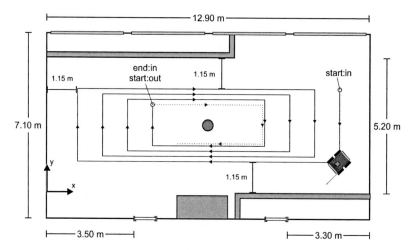

Figure 4.6: *Floor plan of the room that was used as a "two dimensional environment". Apart from a fraction of the rectangular spiral along which sensor readings were collected, the location of the gas source is indicated as well as the windows in the upper part, obstacles in the room, and the doors in the lower part.*

of the container or the type of analyte used. Also the actual sensor position was found to have only a minor influence on the qualitative profile of the obtained sensor readings. The dominant influence of the sensing strategy could be observed in either case. A strong correlation between sensor response and proximity to the gas source was found with neither configuration if a stop-sense-go strategy was applied, while the dominant sensor peaks corresponded approximately with the source location if the readings were collected while the robot was moving at a constant speed of at least 5 cm/s. Moreover, a rough correspondence with the closest approximation to the source could be observed in a "one dimensional scenario" if the source was placed either inside or outside the path of the robot.

4.2   "Two Dimensional" Scenario

It is possible to extend the simple localisation strategy discussed in the previous section in a straightforward manner to a "two dimensional" environment. Instead of searching for local maxima in sensor response along a straight path, readings have to be recorded along a trajectory that covers the designated area. This raises the question of whether local maxima can be observed that correspond to close proximity to a source on a two dimensional search space as well. In order to investigate this question, further experiments were performed in an unventilated room with a rectangular outline (12.9×7.1 m^2). A floor plan of the room is shown in Fig. 4.6. Obstacles are represented by dark shaded regions, and the gas source is indicated by a circle in the middle of the room. The windows (in the upper part of the figure) and the doors (in the lower part of the figure) are also marked. Due to the location of fixed obstacles in the room, the area used for the experiments is bounded by a 12.9×5.2 m^2 rectangle.

In the experiments with the "two dimensional scenario", the robot was driven along a rectangular spiral while keeping a constant velocity of 15 cm/s along the straight lines. The distance between subsequent windings was 1.15 m in x-direction, and 0.20 m in y-direction. Thus, a search strategy is applied that covers the accessible space and takes into account the fact that a strong correlation between sensor response and proximity to a source could only be observed in the "one dimensional" case with a constant velocity. A sketch of the driven path is included in Fig. 4.6 indicating a complete inward spiral (from "start:in" to "end:in") and the beginning of the successive outward movement (starting at "start:out"). Along the straight lines, a control algorithm similar to the one used in the one dimensional case was applied (see Section A.2, especially Fig. A.5) to align the robot with the main axes of the room. In order to minimise self-generated airflow, the 90-degree turns were performed with low speed and the robot was stopped for 10 seconds after every turn. Thus, a complete cycle (including an inward and a subsequent outward movement) lasted about 30 minutes.

4.2.1 Experimental Set-Up

The robot was set-up in the same way for the "two dimensional experiments" as in the experiments in "Corridor–2" (see Fig. 4.1 (b)). The evaluation presented here is confined to the four sensors mounted on the vertical rod. Because they do not show an evident dependency on the height, the average of the four readings is depicted in the illustration of the results below as a robust estimate of the relative concentration at the given location (x,y). Only those values recorded along straight lines were used for the evaluation. The gas source was realised by a container with a cross section of 95 cm^2 filled with ethanol. A trial was started at least 15 minutes after the container had been filled.

4.2.2 Results

Two typical results of the trials in the "two dimensional" scenario are shown in a 3D plot in Fig. 4.7. Thick black lines indicate the outline of the room, while small circles represent the received sensor signal. The location of the source position is marked with a larger circle and the windows in the room are indicated by lightly drawn

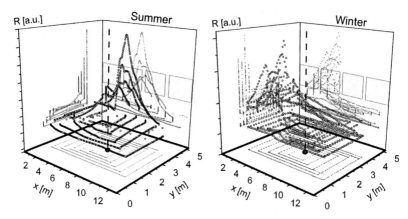

Figure 4.7: *Typical results of the experiments in the "two dimensional" scenario. The graphs show results of measurements performed in the summertime (left) and wintertime (right). During both measurements the source was placed in the middle of the room as indicated by a circle.*

rectangles. A result of an inward run, which was performed in the summertime, is displayed in the left part of Fig. 4.7. As in the "one dimensional scenario", a peak in the profile of the sensor readings was obtained when the robot drove past the source location. Similar peaks, however, occured (with only very few exceptions) always on the side of the source near to the windows. Moreover, the absolute height of the local response maximum increased with increasing distance to the gas source. This can be easily seen on the projection onto the walls in Fig. 4.7. Qualitatively the same results were observed in all three trials performed on three successive days during summertime. The maximal response along the inward path shown in the left of Fig. 4.7 was found at a distance of approximately 1.2 m between the sensors and the gas source. At this point, however, the deviation along the x-axis was only approximately 10 cm. A possible explanation of this observation is the presence of the windows, which act as a heat source on a sunny day. Although the blinds were strictly closed all day long while experiments were carried out, the corresponding side of the room was warmed by the sun whereas the opposite wall remained comparatively cool. In consequence of a stationary heat difference, convection air streams arise, which might prevent increased concentrations in particular regions of the room.

It is therefore expected that a clearly different situation should occur if the heat distribution is reversed as, for example, in wintertime. Here, the wall with the windows was colder than the opposite one, which does not contain a window. The experiments carried out during the winter showed indeed a different picture. A typical result is shown in the right part of Fig. 4.7. In contrast to the summertime, response peaks were found at the wall opposite to the windows. Again, this result was obtained in a similar manner in several trials during two consecutive days.

In fluid dynamics simulations, Krieg demonstrated that under the assumption of similar conditions (an empty room with comparable size and an upwards air current at one wall) the occurence of two cylindric, counterrotating airflow structures is likely [Kri00]. Such room-size coils would circulate around an axis parallel to the x-axis in Fig. 4.7, meaning that the room would be split into two halves with little exchange of gas. The size of the counter-rotating airflow structures is predicted to have a ratio of 2:1. Due to the higher velocity of the ascending air, it can be further speculated that local eddies might occur more often near the windows and that these eddies might account for the higher concentration near to the windows because they can store or accumulate the analyte gas. It has to be emphasised, however, that it is not possible to validate these assumptions based on the experiments discussed in this chapter.

4.3 Conclusions

This chapter presents initial experiments in an environment without ventilation. A first set of trials was carried out in a "one dimensional" scenario, meaning that the robot was driven up and down along a straight line where the gas source was also placed. As an astonishing result, it was found that a strong correlation between sensor response and proximity to a source could be obtained only if the robot was driven with a constant, sufficiently high speed. Such a correlation could not be observed if a stop-sense-go strategy was applied. It has to be mentioned, however, that sensor peaks provide only a rough estimate of the gas source location even if a suitable constant-velocity sensing strategy is applied. While the local concentration maximum in the trials in a "one dimensional" environment was frequently found near the point of closest approximation to the source, deviations of more than 3 m were also observed occasionally. Moreover, the maximal response peak was never found near the point of closest approximation in the experiments in a "two dimensional" environment. There are several reasons that might account for this observation. Assuming that the concentration maximum is shifted from the centre of the gas source by a distance d, a smaller distance $2d/\pi$ is generally expected if the search space is restricted to one dimension. (The proportion of $1{:}2/\pi$ results from integrating over the projections to all possible directions of the search path.) The "one dimensional" scenario also differs from the "two dimensional" one in the physical outline of the room. Due to varying boundary conditions according to the different proportions of length to width, the resulting air flow field is of course expected to be different in the two scenarios. Moreover, it is problematic to compare the scenarios due to the special situation with a strong and approximately stationary heat gradient in the room that was observed in the "two dimensional" environment.

It is therefore difficult to draw universally valid conclusions from the experiments in this chapter. The observations that were made in all the trials, however, can be generalised to some extent just because of the difference between the environments considered. One of these observations is the mentioned impact of the sensing strategy, which was also confirmed independently by Farah [FD02]. Another one is the striking stability of the gas distribution over time, which could be observed in all the trials. A very stable concentration profile was also observed in the majority of the experiments discussed in the remainder of this work, which were carried out in different "two dimensional" environments and also with a different robot and gas sensitive system. This issue is thoroughly discussed in [WLD+03]. A third important insight is that the point of maximum concentration does not usually provide an accurate estimate of the location of the gas source. Local maxima of the instantaneous gas distribution, which can be observed if a stop-sense-go strategy is applied, contain very little information about the source location. An improvement can be achieved if information about the frequency of local maxima is included into the received signal, as is done if the readings are collected at a sufficiently high speed, due to the response characteristics of the metal oxide sensors. Nevertheless, an accurate estimate cannot be guaranteed because the maximum concentration can occur far from the source, especially in the case of a self-contained environment and a gas source that supplies the analyte at a low rate. On the other hand, it is often more important to know where high concentrations of the analyte occur than to know the exact location of the source. The issue of how to combine sensor readings into a single representation of the average gas distribution is addressed in the next chapter.

Chapter 5

Building Concentration Gridmaps

"Sieht fast so aus", meinte Lukas nachdenklich, "als ob hier irgendetwas dazwischen gehört was jemand herausgenommen hat."[1] (Michael Ende, "Jim Knopf und die wilde 13")

5.1 Representation of Gas Distribution Features

This chapter addresses the problem of mapping the structure of a gas distribution from the gas sensor readings collected by a mobile robot. Due to the local nature of single gas sensor measurements (a single metal oxide sensor only provides information about the reactions at the sensor's surface, covering approximately 1 cm^2), it is not possible to measure the entire distribution at the same time without using a dense grid of sensors. With an increasing area, this would involve an arbitrarily high number of fixed gas sensors, which poses problems such as cost and a lack of flexibility. Furthermore, an array of metal oxide sensors would cause a severe disturbance due to the convective flow created by the heaters built into these sensors [ITYM03].

In an environment that is not strongly ventilated, the dispersal of gas tends to be dominated by turbulence and convection flow rather than diffusion, typically resulting in a jagged pattern of temporally fluctuating eddies (see Section 2.2). Due to the chaotic nature of turbulent gas transport, a snapshot at a given instant contains little information about the distribution at another time. It is, however, often sufficient to know the time-averaged properties of a gas distribution. The time-invariant structure of a gas distribution can provide clues that allow a mobile robot to localise a gas source (see Section 5.9). Moreover, it is often desirable to estimate the concentration in a particular area of the environment. In the case of an emergency with a toxic gas, for example, a robot that is able to build a

---

[1] "Looks almost", said Lukas thoughtfully, "as if something is missing in between here, which was taken out by someone." (Personal translation)

map representing the average concentration, could provide valuable information for rational decision making in a rescue mission.

Aside from the task of preparing for rescue missions that have to be accomplished within an environment containing toxic gas, the introduced mapping technique can be useful for any application that benefits from knowledge about the average distribution of a certain gas. For Precision Farming [BG02], gas distribution mapping could be used as a non-intrusive way of assessing certain soil parameters or the status of plant growth to enable a more efficient usage of fertiliser, for example.

To the best of the author's knowledge, there have been only a few suggestions for creating spatial representations of gas distribution, which are discussed in Section 2.4. Unlike the method used by Hayes et al. [HMG02], the algorithm that is introduced in this section does not depend on a perfectly even coverage of the inspected area and can represent fine gradations of the average concentration because not only binary information from the gas sensors is utilised. In contrast to the method by Ishida et al. [ITYM03], data recorded with a mobile robot are used, i.e., there is no grid of gas sensors required.

The gas concentration mapping algorithm introduced in this section produces an estimate of the average relative concentration of a detected gas in a gridmap structure. In order to create such "concentration gridmaps", the gas sensor readings are combined with location estimates provided by another sensor system. Here, the location estimates required for map building were obtained by the external, vision-based absolute positioning system W-CAPS (see Section B). However, the results are expected to apply to any mobile robot equipped with a suitably accurate on-board positioning system, e.g., by carrying out simultaneous localisation and mapping with other sensor systems (see Duckett [Duc03], for example).

In order to obtain complete concentration gridmaps, the path of the robot carrying the gas sensors should roughly cover the entire space, although a uniform exploration is not necessary. As discussed in Section 5.4, it is also advantageous to pass particular points from multiple directions to increase spatial accuracy.

Another assumption, which has to be made, is that the gas distribution in fact exhibits smooth and time-constant structures. This is expected to be usually fulfilled in unventilated and unpopulated indoor environments. As mentioned in Chapter 4 and explicitly discussed in [WLD+03], stable concentration profiles were observed over several hours in a similar scenario for two different indoor environments.

The rest of this chapter is structured as follows. First, the similarities and differences between conventional gridmaps and concentration gridmaps are discussed in Section 5.2. Next, the algorithm to create concentration gridmaps is introduced in Section 5.3. The suitability of the proposed mapping algorithm regarding the slow response and recovery of the gas sensors is then addressed in Section 5.4. After a description of the experimental setup in Section 5.5, the results are discussed with respect to different mapping parameters (Section 5.6) and data acquisition strategies (Section 5.7), the stability of the mapped features over time (Section 5.8) and the capability to use concentration gridmaps to locate a gas source (Section 5.9). Finally, the problem that there was no independent method to verify the actual gas distribution (ground truth) is addressed in Section 5.10 and an iterative formulation of the introduced mapping algorithm for online use is proposed in Section 5.11, followed by conclusions and suggestions for future work (Section 5.12).

5.2 Conventional versus Concentration Gridmaps

Gridmaps were originally introduced to mobile robotics in the early 1980s as a means of creating maps using wide-angle measurements from sonar sensors [EM85; Elf87; BK88; MM96; PNDW98]. The basic idea is to represent the robot's surroundings by a grid of small cells. In a conventional occupancy gridmap, each cell contains a certainty value representing the belief that the corresponding area is occupied by any object.

In the suggested approach, the cells in the gridmap correspond to an estimate of the average relative concentration of a detected gas in that particular area of the environment. There are several problems in creating such a representation that are specific to robots equipped with gas sensors, discussed as follows.

In contrast to range-finder sensors such as sonar or laser, a single measurement from an electronic gas sensor provides information about a very small area because it represents only the reactions at the sensor's surface (≈ 1 cm^2). This problem is further complicated by the fact that the metal-oxide sensors typically used for this purpose do not provide an instantaneous measurement of the gas concentration. Rather, these sensors are affected by a long response time and an even longer recovery time ("memory effect"). The time constants of rise and decay of the Mark III mobile nose used here were estimated as $\tau_r \approx 1.8$ s and $\tau_d \approx 11.1$ s (see Section 3.3). This means that considerable integration of successive measurements is carried out by the sensors themselves. Thus, the output at a given instant depends on the past and the actual magnitude of the stimulus cannot be determined unambiguously from the sensor readings [SLV93]. The impact of this effect on the concentration mapping is discussed in Section 5.4.

To overcome these problems, a mapping technique is introduced that permits integration of gas sensor measurements over an extended period of time. In order to compensate for the small overlap between single measurements, spatial integration of the point measurements is carried out by using a Gaussian density function to extrapolate on the measurements. Thus, a decreasing likelihood is assumed that a given measurement represents the average concentration with respect to the distance from the point of measurement.

By calculating a weighted average of the measurements over time (the weights are determined for each cell with the Gaussian density function), the suggested mapping algorithm models the physical reality inbetween the point locations where

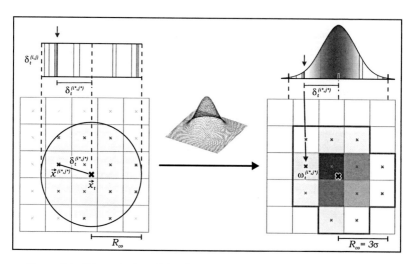

Figure 5.1: *Discretisation of the Gaussian weighting function onto the grid.*

the measurements were taken. Although gas sensor readings represent primarily just
the very small area of the sensor's surface, this technique can be applied, because
the readings contain implicit information about a larger area for two reasons:

- Despite the jagged gas distribution of temporally fluctuating eddies (discussed
  in Section 2.2), it is reasonable to assume that the gas concentration in the
  vicinity of the point of measurement does not change drastically because of
  the smoothness of the time-constant structures in the distribution.

- The metal-oxide gas sensors perform temporal integration of successive readings implicitly due to their slow response and the long recovery time. Thus,
  spatial information is integrated along the path driven by the robot.

Note that errors in measurement are not modelled by the mapping algorithm because
the electronic noise on individual gas sensor readings is neglible [IKNM99]. Rather,
the suggested algorithm models the decreasing amount of information contained in
single sensor readings about the average gas concentration in the vicinity of the
point of measurement. By applying this model, the algorithm is able to determine
the spatial structure of the average gas distribution with limited distortion from
a series of readings, which do not represent the gas concentration directly due to
the response characteristics of the sensors. Individual sensor readings, however, are
assumed to represent the actual physical effect measured (the rate of combustion
reactions) exactly.

5.3 Algorithm to Create Concentration Gridmaps

The sensor readings are convolved using the radially symmetric two dimensional Gaussian function:

$$f(\vec{x}) = \frac{1}{2\pi\sigma^2} e^{-\frac{\vec{x}^2}{2\sigma^2}}. \tag{5.1}$$

This weighting function indicates the likelihood that the measurement represents the average concentration at a given distance from the point of measurement. In detail the following steps are performed (see Fig. 5.1 and Fig. 5.2):

- In the first step the normalised readings r_t are determined from the raw sensor readings R_t as

$$r_t = \frac{R_t - R_{min}}{R_{max} - R_{min}}, \tag{5.2}$$

  using the minimum and maximum (R_{min}, R_{max}) value of a given sensor.

- Then, for each grid cell (i, j) within a cutoff radius R_{co}, around the point $\vec{x}_t$ where the measurement was taken at time t, the displacement $\vec{\delta}_t^{(i,j)}$ to the grid cell's centre $\vec{x}^{(i,j)}$ is calculated as

$$\vec{\delta}_t^{(i,j)} = \vec{x}^{(i,j)} - \vec{x}_t. \tag{5.3}$$

- Now the weighting $w_t^{(i,j)}$ for all the grid cells (i, j) is determined as

$$w_t^{(i,j)} = \begin{cases} f(\vec{\delta}_t^{(i,j)}) & : \quad \delta_t^{(i,j)} \leq R_{co} \\ 0 & : \quad \delta_t^{(i,j)} > R_{co} \end{cases} \tag{5.4}$$

- Next, two temporary values maintained per grid cell are updated with this weighting: the total sum of the weights

$$W_t^{(i,j)} = \sum_{t'}^{t} w_{t'}^{(i,j)}, \tag{5.5}$$

  and the total sum of weighted readings

$$WR_t^{(i,j)} = \sum_{t'}^{t} r_{t'} w_{t'}^{(i,j)}. \tag{5.6}$$

- Finally, if the total sum of the weights $W_t^{(i,j)}$ exceeds the threshold value W_{min}, the value of the grid cell is set to

$$c_t^{(i,j)} = WR_t^{(i,j)} / W_t^{(i,j)} \quad : \quad W_t^{(i,j)} \geq W_{min}. \tag{5.7}$$

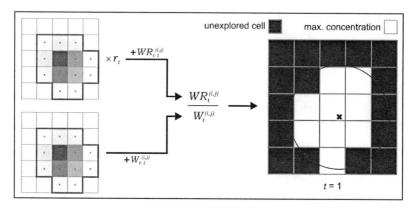

Figure 5.2: *Determining the belief about the average relative concentration.*

An example that shows how a single reading is incorporated into an empty 5×5 map is depicted in Figs. 5.1 and 5.2. First, 13 cells are found to have a distance of less than the cutoff radius from the point of measurement. The cutoff radius is indicated in the left part of Fig. 5.1 by a circle around the point of measurement $\vec{x}_t$, and the affected cells are highlighted in the right part of the figure by a surrounding strong border. Next, the weightings for these cells are determined by evaluating the Gaussian density function for the corresponding displacement values, which are represented by the vertical lines in the upper left part of Fig. 5.1. In the example, the cutoff radius was chosen to be three times the width σ, resulting in the Gaussian shown in the upper right part of Fig. 5.1. The weights are represented by shadings of grey. Darker shadings indicate higher weights, which corresponds to a stronger contribution of the measured value r_t to the average concentration value for a particular cell. The two temporary values maintained per grid cell are illustrated on the left side of Fig. 5.2, symbolising the matrices that contain the sum of the weighted readings $WR_t^{(i,j)}$ (upper part) and the sum of the weigthings $W_t^{(i,j)}$ (lower part). Finally, the grid cells are set to $WR_t^{(i,j)}/W_t^{(i,j)}$ if the sum of the weights exceeds the threshold W_{min}. In the considered example it is assumed that this applies to 9 cells. A concentration map as it might result from integrating a single reading into an empty map (in other words, the map after the first time step) is shown in the right part of Fig. 5.2. Because only one reading is considered, all the cells for which the sum of the weightings exceed the given threshold are assigned with the same relative concentration value r_t. These cells appear in white colour (corresponding to the maximum relative concentration) because they naturally represent the "highest" concentration after the first timestep. The remaining cells are marked as unexplored with a different colour {green}.

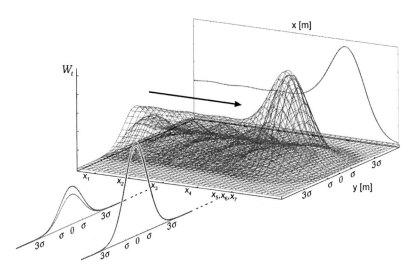

Figure 5.3: *Total weight W_t for an example sensor trajectory {in red}. The trajectory results from a constant velocity movement along a straight path and an immediate stop after the fifth time step (i.e., measurements x_5, x_6 and x_7 were all taken at the same physical location).*

5.3.1 Distribution of the Weighting Function

Given a sequence of measurements, the suggested mapping algorithm determines the total weight W_t according to the path of the sensor. The accumulated weight corresponds to the information about the average concentration that is contained in the measurements recorded along this path. A possible distribution of W_t is shown in Fig. 5.3. To demonstrate the concept of the mapping algorithm, a hypothetical sensor trajectory is considered, consisting of a constant velocity movement along a straight path ($x_1, ..., x_7$, $y_1 = ... = y_7 = 0$) and an immediate stop after the fifth time step ($x_5 = x_6 = x_7$). The driving direction is indicated by a strong arrow.

The 3D plot in Fig. 5.3 shows the Gaussian density function (see Eq. 5.4) for the first four steps, the sum of the Gaussians for the last three steps recorded at the same location and the total sum of all Gaussians on top of it {in red}. It can be seen that the resulting sum W_t models the information content in a way that assigns the strongest weight along the sensor trajectory. Here, the certainty is particularly high because the gas concentration at the corresponding locations actually contributed to the measurements.

Due to the memory effect, the readings of a gas sensor carried by a moving robot represent a spatially and temporally integrated value of the gas concentration that the sensor was exposed to along its trajectory. Consequently, the certainty about the average concentration is modelled as being higher if the robot is driven at a slower speed. A strong weight is applied by the mapping algorithm especially in cases where a number of successive measurements were performed at a particular spot (as in the last three time steps in the example shown in Fig. 5.3). Here, the assigned concentration value (calculated by averaging the three recorded readings) represents a temporally integrated quantity that naturally contains more information about the average concentration at this particular location. A temporally integrated value also contains more information about adjacent places because of the higher certainty about the average concentration, which is extrapolated by assuming a smooth transition. The temporal mean also carries out some spatial integration due to the spatial fluctuation of the gas. A comparison of the total weight, assigned to a location the robot passed by at a constant speed, with the weight, determined for the location where the three successive measurements were taken, is depicted in the lower left part of Fig. 5.3 by two projections at $x = x_3$ and $x = x_5, x_6, x_7$ respectively.

While the mapping algorithm assigns a strong weight along the trajectory, the information content decreases quickly orthogonal to the path. Thus, the mapping algorithm takes into account the fact that the measurements were actually not performed at the corresponding locations and the concentration values can be estimated only by extrapolating on the actual measurements assuming smooth transitions in the time-constant gas distribution structure. Note that the mapping algorithm assigns a quickly decreasing weight to the front end of the path in the same way as orthogonal to it.

Another feature of the resulting weighting distribution can be seen in the projection along the sensor path ($y = 0$), shown in Fig. 5.3 behind the 3D plot: the weight stays approximately constant along the path in case of an equidistant spacing between the locations where the measurements were performed. Because they represent an integrated value, the information content of the sensor readings about locations along the integration path is in fact approximately independent of the actual points of measurements, if the time constant of decay τ_d is much longer than the time between individual measurements Δt_r. Consequently, the certainty about the average concentration at a location in between two points of measurement (calculated by averaging over adjacent measurements), can be comparably high as the value directly at a point of measurement.

As shown in Appendix C.1 on the basis of an infinite series of equidistant point measurements, W_t can be considered as constant along the path if the step width Δx is smaller than the width of the Gaussian multiplied by a constant C_{x_W}. For practical considerations a value of $C_{x_W} = 2$ is usually sufficient, which limits the variation along the path to 2.9 %. Using Eq. C.11 and the linear speed of the robot v, this can be expressed as

$$\sigma > \frac{\Delta x}{C_{x_W}} = \frac{v \Delta t_r}{C_{x_W}} = \sigma_{min}. \tag{5.8}$$

In the experiments presented below, the maximum velocity was set to 5 cm/s and the measurement frequency was approximately 1.25 Hz, resulting in a lower limit of $\sigma_{min} = 2$ cm. Note that for the sake of clarity a step width of $v \Delta t_r = 2\sigma$ was chosen in Fig. 5.3, which corresponds to the lower limit of $\sigma = \sigma_{min}$. At the same time, the above mentioned condition, that the time constant of decay must be longer than the time between the measurements, has to be fullfilled:

$$\tau_d \gg \Delta t_r. \tag{5.9}$$

This condition is satisfied in the experiments presented below because $\tau_d \approx 11.1$ s and $\Delta t_r \approx 0.8$ s.

While the information content along the path is modelled approximately independently of the chosen parameter σ within the scope defined by the Equations 5.8 and 5.9, the relative width orthogonal to the path is determined by this parameter only (see Appendix C.2). The width of the Gaussian weighting function therefore provides a parameter to adjust the amount of extrapolation that is carried out by the mapping algorithm over areas where no measurements were taken. It has to be set according to the driven path by balancing the need for sufficient overlap of the measurements and the desire to preserve small structures in the gas distribution. A detailed discussion of the choice of mapping parameters is given in Section 5.6.

5.4 Impact of the Sensor Dynamics

Due to the response characteristics of metal oxide sensors, a single gas sensor reading represents a temporally and, if the robot is driven at non-zero speed, also a spatially integrated concentration value. The averaging effect is considered implicitly by the model of the information content applied in the mapping algorithm. The weighting function (Eq. 5.1) that represents the information content, contains, on the other hand, no term to model the asymmetry, which is induced by the long response and recovery time of the sensors. It would indeed not be possible to unambiguously determine the actual concentration distribution along the path that caused a given series of gas sensor readings. In order to avoid extensive analysis of the temporal profile of the readings and the driven trajectory, it was decided to omit an asymmetric term in the Gaussian weighting function (Eq. 5.1). Consequently, a certain level of distortion in the mapped gas distribution has to be tolerated. The magnitude of this distortion is estimated below.

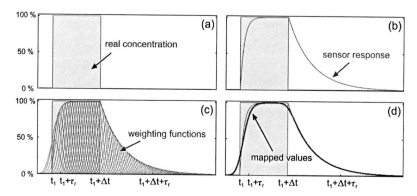

Figure 5.4: *Mapping of a rectangular step pulse. The figure shows the step-like concentration course the gas-sensitive system is exposed to (a), the sensor response as calculated for the Örebro Mark III mobile nose (b), the Gaussian weighting functions multiplied by the corresponding sensor readings (c) and the resulting curve of the mapped values (d).*

As a consequence of the delayed response and the prolonged decay time of the gas sensors, the mapped values show asymmetrically blurred edges and a slightly shifted centre of the area of maximum concentration compared to the real distribution. This effect can be seen in Fig. 5.4, which shows how a rectangular step pulse would be mapped by the gas concentration mapping algorithm introduced in Section 5.3. In the upper left part (a) the real distribution can be seen, which is a step pulse with an assumed duration of $\Delta t = 10$ s. In addition, the response of the gas-sensitive system is shown in part (b). This curve was calculated using a first order sensor model and the parameters $\tau_r \approx 1.8$ s and $\tau_d \approx 11.1$ s of the Örebro Mark III mobile nose (see Section 3.3). In Fig. 5.4 (c), the Gaussian weighting functions multiplied by the corresponding sensor readings are also shown. The samples were assumed to be recorded at a rate of 2 Hz and a width of $\sigma = 1$ s was used for the Gaussian weighting function (see Eq. 5.1). This corresponds to a distance of $\sigma = 5$ cm, if a situation is considered where a robot drives with a constant velocity of 5 cm/s through a 50 cm wide area of constant concentration. Note that the Gaussians vanish in the front part of the graph due to the zero response of the sensor. Finally, the normalised curve of the mapped values is depicted in Fig. 5.4 (d). This curve is calculated according to Eq. 5.7, meaning that the sum of the Gaussians shown in Fig. 5.4 (c) is divided by the sum of the weights, which is approximately constant because the conditions given by Eqs. 5.8 and 5.9 are fullfilled for the example considered.

Comparing the real distribution with the course of the mapped values, the asymmetrical shift as well as the blurring effect can be seen in Fig. 5.4 (d). This corruption

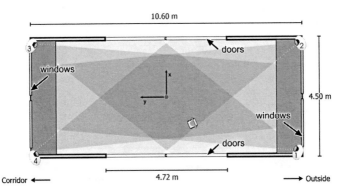

Figure 5.5: *Laboratory room in which the gas concentration mapping experiments were performed. Indicated are the web-cameras that were used to track the position of the robot (see Appendix B), and the location of doors and windows.*

is, however, not critical. Due to the low speed of the robot, which never exceeded 5 cm/s during the experiments presented here, the expected shift would be in the order of 10 cm at most. This effect is even smaller for smooth distributions, which the metal-oxide sensors can follow more closely than a step-like one. Further on, the directional component of both effects would be averaged out if the robot passed the same point from different directions. If this condition is fulfilled, the position of concentration maxima is represented closely by the described mapping process. The remaining effect of the blurred edges should be tolerable as long as the time-constant structures can be distinguished in the maps created.

5.5 Experimental Setup

The experiments to test the concentration mapping algorithm were performed in a rectangular laboratory room at Örebro University (size 10.6×4.5 m²). A floor plan of this room is shown in Fig. 5.5. The air conditioning system in the room was deactivated in order to eliminate the possibility of a dominant constant airflow.

The testing platform was chosen to be a Koala mobile robot equipped with the Mark III mobile nose. This comparatively small robot is described in Section 3.2.1, and the Mark III mobile nose is detailed in Section 3.2.2. Fig. 3.1 shows the robot carrying the mobile nose and a cardboard hat that is used by the absolute positioning system W-CAPS to track the robot's position (see Appendix B for details of the positioning system).

To emulate a typical task for an inspection robot, the gas source was chosen to imitate a leaking tank. This was realised by placing a paper cup filled with ethanol on a support in a bowl with a perimeter of 12 cm (see Fig. 3.1). The ethanol dripped through a hole in the cup into the bowl at a rate of approximately 50 ml/h. Ethanol was used because it is non-toxic and easily detectable by the metal oxide sensors.

5.6   Mapping Parameters

The concentration mapping algorithm requires three parameters: the width of the Gaussian σ, the cutoff radius R_{co} and the threshold W_{min}. While the actual values of R_{co} and W_{min} do not have a strong influence on the resulting concentration map, the parameter σ is crucial. As mentioned above, a value of σ has to be chosen that is high enough to satisfy the requirement for sufficient extrapolation, but low enough to preserve the fine details of the mapped structures. A comparison of concentration gridmaps that were created using different values of σ is shown in Fig. 5.6. These maps were created from the same sensor data collected over a period of 60 minutes with two pairs of equivalent sensors (TGS 2600 and TGS 2620) mounted symmetrically on both sides of the robot. In the examples, a cell size of 2.5×2.5 cm^2, a cutoff radius of $R_{co} = 3\sigma$ and a weighting threshold of $W_{min} = 1.0 \times$ (number of sensors) $= 4.0$ were used. In order to acquire sensor data, the robot was driven along a predefined path given by a sequence of rectangular spirals centred at the position of the gas source (see Section 5.7.1 for details). The position of the gas source is indicated in Fig. 5.6 by a hollow circle.

All the concentration gridmaps presented in this work are illustrated like those in Fig. 5.6. Different shadings of grey are used to indicate concentration values whereas dark shadings correspond to low and light shadings to high relative concentrations. The values higher than 90% of the maximum are indicated with a second range of dark-to-light shadings {of red}. Unexplored cells are displayed with another colour {green}. Note that the chosen illustration shows average concentration values relative to the dynamic range in the map. The values in a concentration gridmap were determined from the relative sensor readings, which by defintion lie in the range of [0,1]. Because the calculated range of average concentration values is usually narrower, a linear transformation is thus applied in order to display the structure of the gas distribution using the full range of available shadings. Consequently, the absolute value of a grid cell in two different gridmaps cannot be compared directly by comparing the corresponding brightness levels.

If a narrow weighting function is applied, the resulting gridmaps are dominated by the local variations that occured along the path of the robot. As indicated by the gridmap shown in the upper left part of Fig. 5.6, this can be seen especially in case of very low values of σ where the map basically represents the sequence of readings along the path driven. Increasing σ causes local maxima to be combined,

90%

50%

σ = 1 cm, R_{co} = 3 cm $\qquad$ σ = 5 cm, R_{co} = 15 cm $\qquad$ σ = 10 cm, R_{co} = 30 cm

1m

σ = 15 cm, R_{co} = 45 cm $\qquad$ σ = 20 cm, R_{co} = 60 cm $\qquad$ σ = 30 cm, R_{co} = 90 cm

Figure 5.6: *Comparison of concentration gridmaps created from the same data set using a Gaussian weighting function with different width σ. The data set was acquired while the robot was driven along a sequence of rectangular spirals centred at the position of the gas source (indicated by a circle).*

and thus larger structures of the gas distribution to appear as contiguous patches. This applies of course to local maxima that appear in consequence of insufficient overlap as well as to small variations that represent actual time-constant structures of the gas distribution. Consequently, only the large scale structures are mapped in the case of very wide weighting functions. In the gridmap shown in the lower right part of Fig. 5.6 (σ = 30 cm), for example, only a large plume-like structure remains indicating a weak airflow towards the left side (west). As a rather small Gaussian width, that nevertheless permits averaging of local variations, a value of σ = 15 cm was found to be suitable regarding the trajectories in the experiments presented in this chapter. All of the subsequently presented gridmaps were created using this parameter value and a cutoff radius of $R_{co} = 3\sigma$.

The comparison in Fig. 5.7 shows that the basic structure of the concentration gridmaps is not affected by the choice of the threshold W_{min}. In the case of very low threshold values, however, grid cells at the border can be assigned with unreliable values because only a few or even just one measurement might have been

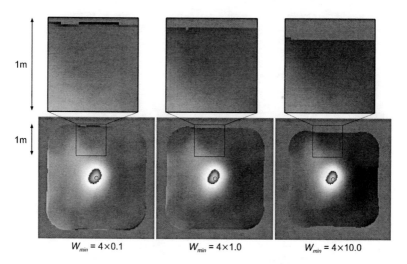

Figure 5.7: *Comparison of concentration gridmaps for different weighting thresholds W_{min}. The maps were created from the same data set that was used for Fig. 5.6. Concentration values higher than 90% of the maximum are indicated with a second range of dark-to-light shadings {of red}.*

considered for averaging. This can be seen in the left part of Fig. 5.7. The threshold $W_{min} = 4 \times 0.1$ corresponds to the relative value of the Gaussian weighting function at a distance of $\approx 2.15\ \sigma$. Considering only one sensor, a single reading would be sufficient to establish the value of grid cells within this distance from the point of measurement. Consequently, the cells at the border of an inspected area are prone to be assigned with extreme values, which are calculated based on only a few readings recorded at outer positions. Note that the extremely low concentration values, which are included at the border of the map in the left part of Fig. 5.7, lead to a reduced minimal concentration value in the map. As a consequence, the map appears brighter compared to the maps shown in the middle and the right part of this figure.

Using a parameter value of $W_{min} = 4 \times 1.0$ (as in the middle part of Fig. 5.7), it is guaranteed that the grid cell values are obtained by averaging over more than one sensor reading. (Note that while the maximum of the relative value of the Gaussian weighting function is 1.0, a sum of 4.0 cannot be obtained for one measurement for a particular cell due to the different locations of the gas sensors.) With an increasing value of W_{min}, the area that is represented in the map becomes slightly smaller because more and more grid cells at the border of the inspected area are considered as not sufficiently examined (unexplored). At the same time, abrupt transitions at the border vanish. Thus, the parameter W_{min} has to be set to a sufficiently high value in order to avoid unreliable cells at the border of the inspected area, while its exact value is not crucial. For the remainder of this work, a value of $W_{min} = 10.0 \times$ (number of sensors) was used.

5.7 Data Acquisition Strategy

The introduced mapping technique allows to separate the underlying structure of a gas distribution from the transient variations by integrating many measurements along the path of the robot. It was mentioned above that the actual trajectory along which the sensor readings were recorded is expected to have a minor influence on the resulting concentration map. While uniform exploration is not necessary, it is required that the trajectory roughly covers the available space. In order to obtain better accuracy, it is further advantageous to pass particular points from multiple directions (see Section 5.4). This can be achieved in a straightforward way by driving the robot along a *predefined path* that complies with the mentioned requirements. Two different predefined trajectories, a rectangular spiral and a sweeping movement (see Sections 5.7.1 and 5.7.2) were tested here. On the other hand, it might be not desirable to specify a fixed trajectory if a robot is not exclusively used for gas distribution mapping. Instead, it might be very useful for some applications (e.g., search and rescue operations) if gas sensors could easily be added to existing mobile robots without the need to severely modify the behaviour. Especially in the case of rescue or surveillance robots that are intended for exploration and searching tasks, this could provide a valuable upgrade of the capabilities of the robot. The same applies naturally if the target of a searching task is a gas source. Consequently, the concentration mapping algorithm was also tested with data obtained using *reactive gas source tracing* strategies (see Section 5.7.3).

5.7.1 Predefined Path – Rectangular Spiral

For a first set of experiments, the robot was driven along a sequence of rectangular spirals around the location of the gas source. This path is sketched in the left part of Fig. 5.8 (a). It consists of a sequence of inward and outward movements with a minimal distance to the centre of the source of 1 m, 0.75 m, 0.5 m, 0.35 m on the subsequent windings of the path. After an inward spiral is completed, the robot turns by 180° and drives back to the starting point of the inward spiral along the same path. Thus, points are passed equally often from opposite directions along a trajectory that covers the inspected area. In order to reduce self-induced disturbance of the gas distribution, the robot was driven at a constant speed of 5 cm/s along the straight lines. As discussed in Chapter 4, a constant speed was found to enhance the

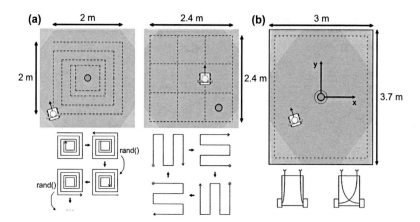

Figure 5.8: *Different data acquisition strategies performed for concentration mapping. (a) Outline of the predefined paths tested. (b) Sketch of the setup used for reactive gas source tracing experiments. The icons indicate the two kinds of Braitenberg vehicles tested in these experiments, either with crossed or uncrossed sensor-motor connections (see Section 6.2 for details).*

gas source localisation capability in a one dimensional environment. At the corners, the robot was rotated slowly ($10°/s$) in order to minimise additional disturbance. A complete cycle including an inward and an outward phase lasted about 25 minutes. These cycles were repeated with a randomly chosen starting corner and direction at the start of each trial. Note that the inspected area is not explored uniformly because of the need to avoid collisions with the beaker in the centre.

5.7.2   Predefined Path – Sweeping Movement

In a second set of experiments, a different predefined path was applied, which is shown in the right part of Fig. 5.8 (a). Here, the robot performs a sweeping movement that encloses nine squares (80 cm × 80 cm) providing nine possible locations in which to place the gas source (one is indicated in the figure). The sweeping movement was implemented as a trajectory consisting of the four segments shown in Fig. 5.8, which were executted repeatedly in the given sequence. Again, a constant speed of 5 cm/s was applied along the straight lines and the robot was rotated with a speed of $10°/s$ at the corners. In contrast to the rectangular spiral path defined in the previous section, points on the path are not traversed equally often from opposite directions. Opposite edges of the nine squares, however, are passed from opposite directions in the course of each cycle. The trajectory also does not provide completely uniform exploration, because the outer edge of the explored area is traversed more often than the inner parts. In contrast to the spiral movement, however, the location of the source is not especially distinguished by the symmetry of the path.

5.7.3 Reactive Gas Source Tracing

The gas source tracing strategies tested were realised in the manner of a Braitenberg vehicle [Bra84], meaning that, based on the stereo architecture of the mobile nose, a direct sensor-motor coupling was implemented. Details of the implementation and a discussion of the possible use of such reactive strategies to the problem of gas source localisation are given in Chapter 6. Here, data collected with two Braitenberg-type strategies implemented with uncrossed and crossed sensor-motor connections were used for concentration mapping. The experiments were carried out in a restricted area whereas the boundaries were realised by assigning a repulsive artificial potential field [Kha85] that starts to be effective at a distance of 20 cm from the border. Both, the inspected area and the border region where the repulsive potential is effective are indicated in Fig. 5.8 (b). Because inhibitory connections were used, maximum wheel speed results if the sensed concentration is low, which in turn implements a simple sort of exploration behaviour. Thus, the robot turns towards higher concentrations with uncrossed connections ("exploration and hill climbing" also called "permanent love" by Braitenberg) while the robot turns away from them with crossed connections ("exploration and concentration peak avoidance" also called "exploring love").

In contrast to the experiments where the robot was driven along a predefined path, the trajectory is of course not known in advance in the case of Braitenberg-type strategies. It is thus not guaranteed that the path complies with the mentioned conditions for concentration mapping. In reaction to a certain gas distribution, the resulting trajectory might not cover all regions of the assigned area and the robot might always approach some points from the same direction. It is also known that the "bouncing" behaviour of the robot due to the artificial potential fields (i.e., specular reflection) does not guarantee uniform coverage of an environment [Gag93]. Thus, it can be expected that the area under investigation is usually not uniformly explored.

5.8   Stability of the Mapped Structures

Due to the local character of gas sensor measurements, it takes some time to build concentration gridmaps. In addition to spatial coverage, a certain amount of temporal averaging is also necessary to represent the time-constant structure of the gas distribution. The evolution of time-constant structures can be seen in Figs. 5.9 –

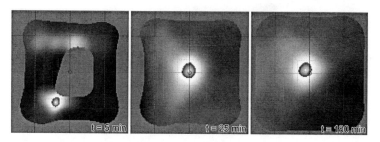

Figure 5.9: *Evolution of concentration gridmaps created from data recorded along a rectangular spiral path (experiment "RectSpiral-4", see also Fig. D.2).*

5.11, which show sequences of gridmaps created from the data collected during a single trial up to the specified time. Additional snapshots from all the experiments performed are given in Appendix D. The gridmaps were created with a cell size of 2.5×2.5 cm^2 using two pairs of equivalent sensors (TGS 2600 and TGS 2620) and the parameter set $\sigma = 15$ cm, $R_{co} = 3\sigma$, $W_{min} = 4 \times 10.0$. The maximal grid cell is indicated in the concentration maps by a small {blue} dot in the lightest part of the second range of dark-to-light shadings {of red}. The 90% median (the median of the x- and y- coordinates of the area, which is defined by those cells with a value of at least 90% of the maximum) is indicated by a second small dot with a different color {light green}.

5.8.1 Predefined Path

Fig. 5.9 shows three snapshots of a concentration gridmap created from data recorded while the robot was driven along a rectangular spiral as defined in Section 5.7.1. Because the area was not yet fully explored after five minutes, an incomplete map was obtained that exhibits some temporary structures of the gas distribution. These structures changed quickly while new readings were added. After 25 minutes, however, a fairly stable representation was attained in all four experiments of this type. A time of 25 minutes corresponds to one completed cycle including an inward and an outward phase, meaning that each point along the spiral path was passed two times from opposite directions. Selected snapshots of the concentration gridmaps for all experiments where the robot was driven along a spiral path are shown in Appendix D.1. The stabilisation of the mapped gas distribution is also indicated by the stabilisation of the distance between the centre of the gas source and the centre of the grid cell with the maximum value, which is shown in Fig. 5.12 (top) for the rectangular spiral experiments. The distance between the maximal cell and the source converged after 20 minutes to a comparatively stable value.

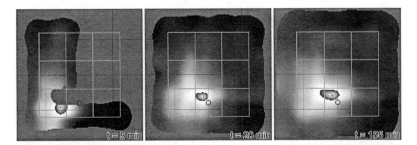

Figure 5.10: *Evolution of concentration gridmaps created from data recorded along a predefined sweeping path (experiment "Sweeping-1", see also Fig. D.3).*

After the first cycle of the spiral path has been completed, only minor changes in the concentration gridmap could be observed, indicating a stable gas distribution profile over time. Occasionally a slight rotation of the direction of the plume-like structure appeared as in the experiment "RectSpiral-4" shown in Fig. 5.9, for example. Here, the plume-like structure was directed westwards after 25 minutes and the direction changed to south-west until the end of the trial.

A similar trend could be observed in the experiments where a sweeping movement was executed by the robot as defined in Section 5.7.2. Corresponding snapshots are shown in Appendix D.2. One example is singled out in Fig. 5.10. Again, the stabilisation of the mapped structures occured within the first complete cycle, after which each point along the path was passed at least two times from opposite directions. The distance of the maximal cell to the gas source is shown in Fig. 5.12 (bottom) for the sweeping experiments. It exhibits considerable step-like variations in two trials ("Sweeping-5" and "Sweeping-8") where the location of the maximal cell changed within an otherwise time-constant structure after a stable location of the maximal cell had been established (see Appendix D.2).

It takes approximately 25 minutes to complete a sweeping cycle comprising the four parts shown in Fig. 5.8. Stable structures in the concentration map, however, could be observed more quickly in all the experiments of this type. A stable representation was achieved after approximately 10 minutes of exploration along the sweeping path, which is faster than the spiral experiments, especially considering the fact that the inspected area was approximately 35 % larger in the sweeping trials. It is likely that this is due to the fact that locations inside the lattice, which is formed by the sweeping path, were passed *nearby* two times from opposite directions after the first part of the sweeping movement had been completed, corresponding to 8.5 minutes of exploration. The path itself, however, was not traversed from opposite directions by the chosen sweeping movement. Instead, the robot was driven along the long tracks always in the same direction. In consequence, the gridmap represen-

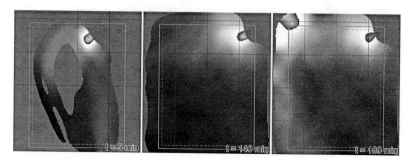

Figure 5.11: *Evolution of concentration gridmaps created from data recorded in a reactive gas source tracing experiment where an "exploration and hillclimbing" strategy was applied (experiment "PL-UR1", see also Fig. D.7).*

tation obtained with a Gaussian width of $\sigma = 15$ cm exhibits cords of high intensity along the driven path, drawn-out from areas of high concentration (see Fig. 5.10 and Appendix D.2, respectively). As a result of the memory effect of the metal oxide sensors, these cords appear due to an insufficient overlap of the measurements regarding the trajectory of the robot. Hence, they disappear if the applied Gaussian is broadened.

5.8.2   Reactive Strategies

Slightly different results were obtained when the robot was controlled reactively as a Braitenberg vehicle (defined in Section 5.7.3). Figs. 5.11 and 5.15 show snapshots of a concentration gridmap that was created from data recorded while the robot was controlled as a Braitenberg vehicle with uncrossed connections ("permanent love"). In the corresponding experiments, a larger area could be inspected, because the reactive control scheme is not very susceptible to occasional dropouts of the information from the external positioning system that might occur in outer regions. It would be otherwise problematic to control a predefined trajectory there if the robot had to rely on its odometry only without external correction. Considering the designated area exclusive of the part where the repulsive potential (see Fig. 5.8 (b)) is effective, the enlargement is approximately 40 % compared to the sweeping trials and approximately 90 % compared to the spiralling experiments. In addition, the average speed was lower even if a speed gain of $K_v = 5$ cm/s was applied, because of the deceleration in response to an elevated gas concentration.

The course of the distance between the maximal cell and the centre of the gas source, which is shown in Fig. 5.13, exhibits nevertheless only a short initial range of abrupt and drastic variation. Preliminary structures were established quickly by

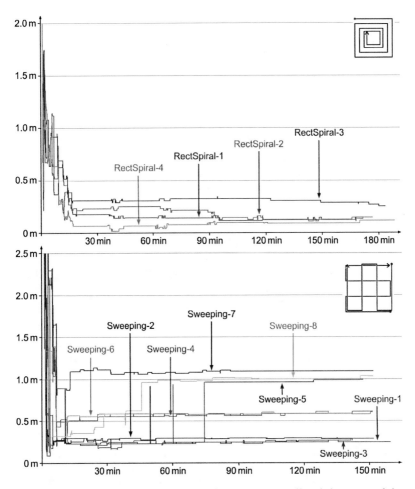

Figure 5.12: *Distance between the maximal concentration cell and the centre of the gas source in the experiments where a predefined path was driven. Top: rectangular spiral as defined in Section 5.7.1. Bottom: sweeping movement as defined in Section 5.7.2. Snapshots of the corresponding concentration maps are shown in Appendix D.1 (rectangular spiral path) and in Appendix D.2 (sweeping movement).*

the robot due to the fact that areas with high concentration were actively explored. By contrast to the experiments with a predefined path, the location of the detected maximum changed more often because the robot has a tendency to get stuck in local concentration maxima. Further, the maximum was not always shifted within an otherwise stable structure but rather the shift was caused by a considerable change in the mapped distribution sometimes. Though the sudden and strong shift of the maximal cell at the end of trial "PL-UR1" is surely the most impressive example for it, it is not representative. Only in this experiment, an abrupt relocation of the centre of the mapped distribution could be observed that is likely to be caused by a relocation of the real distribution. As can be seen in the snapshots of the corresponding trial in Figs. 5.11 and D.7, a stable representation was obtained in the first part of the experiment. The mapped distribution indicated a plume-like structure originating from a point nearby the source in the upper right corner of the inspected area. After 140 minutes, however, a second region of high gas concentration appeared in the upper left corner. In the following, this region was intensely explored by the reactively steered robot (the trajectory is depicted in Fig. E.11) and thus a new region of highest concentration appeared in the map. In all the other trials, abrupt changes of the location of the maximum concentration grid cell occured within a relatively stable structure or due to exploration of a beforehand unexplored area. Snapshots of the concentration maps created from the reactive gas source tracing data are shown in Appendix D.3.1 ("exploration and hillclimbing") and Appendix D.3.2 ("exploration and concentration peak avoidance"), respectively.

5.9   Suitability for Gas Source Localisation

In the case of a gas distribution controlled purely by diffusion, the location of a static, constantly evaporating gas source would correspond to the concentration maximum in the gridmap. As mentioned in Section 2.2, this assumption is not fulfilled under realistic conditions due to the relatively slow diffusion velocity of gases compared to spreading by turbulence and convection flow. However, for the indoor environment investigated, it was found that the concentration maximum could provide an approximate estimate of the source location in many cases, as can be seen in Figs. 5.12 – 5.13. Here, the distance between such an estimate and the centre of the real position of the gas source is plotted against time, for data collected while the robot was driven along a predefined path (see Fig. 5.12) and with a Braitenberg-type strategy (see Fig. 5.13). An exact agreement with the position of the gas source, corresponding to a distance of below 6 cm, was observed only temporarily. On the other hand the error at the end of a trial was less than 75 cm in 16 out of 23 and less than 50 cm in 11 out of 23 experiments. It is not guaranteed, however, to derive a good estimate of the source location from the position of the maximum in a concentration gridmap.

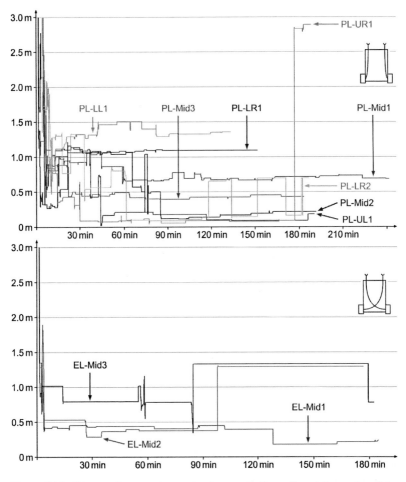

Figure 5.13: *Distance between the maximal concentration cell and the centre of the gas source. The sensor readings were collected while the robot executed a reactive gas source tracing strategy as defined in Section 5.7.3. Top: permanent love (PL, Braitenberg vehicle with uncrossed sensor motor connections). Bottom: exploring love (EL, Braitenberg vehicle with crossed sensor motor connections). The suffix in the identifier of each experiment specifies the position of the gas source (LL corresponds to a gas source in the lower left corner, for example). Snapshots of the corresponding concentration maps are shown in Appendix D.3.*

Type	< 0.25 m	< 0.5 m	< 0.75 m	< 1 m	< 1.5 m	> 68.3%
RectSpiral	74.7 %	100.0 %	100.0 %	100.0 %	100.0 %	≤ 21.5 cm
Sweeping	23.5 %	42.7 %	69.0 %	80.0 %	100.0 %	≤ 57.0 cm
Braitenberg, PL	37.9 %	49.8 %	69.9 %	72.8 %	97.5 %	≤ 74.0 cm
Braitenberg, EL	11.3 %	45.2 %	45.7 %	58.2 %	100.0 %	≤ 130.0 cm

Table 5.1: *Percentage of time steps after an initial stabilisation period of 30 minutes, for which the distance between the maximum concentration grid cell and the centre of the gas source was below the specified thresholds. The itemised statistical values were calculated from 4 spiralling trials with a total length of 725 minutes, 8 sweeping trials (1216 minutes), and 11 Braitenberg trials (2050 minutes). The covered experimental area was approximately 2.4 × 2.4 m² (Spiral), 2.8 × 2.8 m² (Sweeping), and 3.4 × 4.1 m² (Braitenberg).*

A more detailed analysis is presented in Figure 5.14 and Table 5.1. In the figure, the distribution of the distance between the maximal cell and the source is depicted by means of histograms with a bin size of 5 cm. This analysis is intended to show the localisation capability of the gridmaps after stabilisation, so the gridmaps obtained before the assumed stabilisation time were omitted from the calculation. The stabilisation time was assumed to be 30 minutes. Table 5.1 gives a summary of the results obtained in all trials, giving the percentage of time steps for which the distance from the source to the concentration maximum fell below some selected thresholds. The rightmost column of the table gives the radius of the circle around the maximum concentration cell that contains the centre of the gas source in 68.3% of the cases. A value of 68.3% corresponds to the expected fraction of counts of a normally distributed random variable with a distance lower than the width of the distribution σ. Although the observed gas distribution is obviously not normal, this value provides an indicator of the localisation accuracy obtained by the different strategies.

The best performance was obtained with the spiralling strategy, where the 68.3 % radius was found to be 21.5 cm, which is a fairly good result compared to the maximum distance of the sensors to the source of approximately 1.6 m in these experiments[2]. A lower accuracy was observed in the sweeping experiments. The 68.3 % radius was 57 cm, while the maximum distance of the sensors to the source was 1.9 m. This could be explained by the fact that points are not traversed equally often from opposite directions, as discussed in Section 5.4. The same applies to the reactive Braitenberg-type strategies, which are also prone to uneven coverage of the experimental area due to their tendency to get trapped by local concentration maxima. A 68.3 % radius of 74 cm (permanent love) and 130 cm (exploring love)

---

[2]The maximum distance of the gas sensors is reached while the robot is rotating in a corner of the outer winding of the path. Considering a separation between the sensors and the centre of the robot of 20 cm, the distance is given by $(\sqrt{2} + 0.2)$ m = 1.61 m.

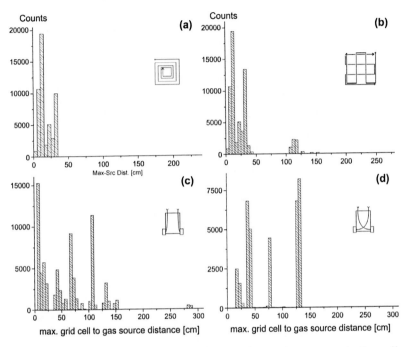

Figure 5.14: *Histogram of the distance between the maximum concentration cell and the gas source for different groups of experiments. Distance values obtained in the initial stabilisation phase of each trial were not considered for calculation. The stabilisation time was assumed to be 30 minutes. (a) Rectangular spiral, (b) sweeping movement, (c) Braitenberg vehicle with uncrossed connections (permanent love), (d) Braitenberg vehicle with crossed connections (exploring love).*

was found, while the maximum distance was 2.3 m in the trials where the source was placed in the middle and 4.55 m in the trials where the source was placed in a corner of the test area. As can be seen in Table 5.2, a slight improvement of these results can be obtained if the 90 % median is used to approximate the location of the gas source. This applies especially in cases where a preliminary location of the concentration maximum is frequently found near the border of the incompletely mapped area, as it often occured in the reactive strategies tested.

The available experimental results suggest that it is possible to determine the quality of an estimate from the shape of the mapped distribution. A sharply peaked representation with an approximately circular region of highest intensity yielded the

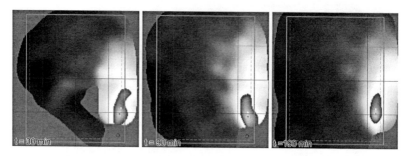

Figure 5.15: *Example of a trial where a plume-like area of high concentration was observed. The concentration gridmap was created from data recorded in the reactive gas source tracing experiment "PL-LR1" where an "exploration and hillclimbing" strategy was applied.*

best localisation results (see Appendix D). If the region that contains values larger than 90 % of the maximum developed a more plume-like shape, the position of the maximum in the concentration gridmap was less suitable to approximate the location of the source. And the worst localisation performance was obtained in cases where a large plume-like structure was found, which is stretched out up to the edge of the mapped area, while it is broadened towards the border and has a maximum that is located at the edge (see the experiments "Sweeping-5", "Sweeping-8" and "PL-UR1" in Appendix D, for example). Finally, a poor localisation was also observed in case of a disjointed region of highest intensity as in the trials and "EL-Mid3" and "PL-Mid3".

This correlation points to an important issue in the context of localising a gas source from a gas distribution map. Depending on the actual airflow situation, it *can* be meaningful to approximate the location of a gas source by the location of the maximum average concentration. This applies to conditions where a sharply

Type	< 0.25 m	< 0.5 m	< 0.75 m	< 1 m	< 1.5 m	> 68.3%
RectSpiral	73.2 %	100.0 %	100.0 %	100.0 %	100.0 %	≤ 21.5 cm
Sweeping	3.5 %	38.2 %	77.0 %	87.5 %	100.0 %	≤ 69.0 cm
Braitenberg, PL	18.4 %	45.4 %	71.2 %	74.1 %	98.1 %	≤ 70.0 cm
Braitenberg, EL	1.2 %	45.2 %	75.7 %	99.9 %	100.0 %	≤ 62.0 cm

Table 5.2: *Percentage of time steps after an initial stabilisation period of 30 minutes, for which the distance between the 90% median grid cell and the centre of the gas source was below the specified thresholds.*

peaked distribution with an approximately circular region of highest intensity was found, probably indicating a diffusion-like spreading of gas. In these cases, the best localisation results were obtained. On the other hand, the location of maximal concentration does *not* necessarily provide a reasonable estimate of the location of a gas source. Poorer localisation performance was obtained when the area of high concentration formed a plume-like structure, as can be seen in Fig. 5.15, for example. Such a plume-like structure might indicate a stronger airflow, which would tend to increase the distance of the concentration maximum from the gas source. A further possible explanation is the existence of local eddies that are able to "store" the analyte gas. The location of such eddies and thus the location of the corresponding maximum is not directly correlated with the location of the source. Finally, the maximal average concentration is also expected not to be suitable for gas source localisation if the airflow situation changes drastically during an experiment. Consequently, the gas source localisation statistics given in Figures 5.12 – 5.14 and Tables 5.1 – 5.2 have to be treated with care because they summarise the results of trials in which different airflow situations were encountered.

Nevertheless, it is possible to achieve good gas source localisation performance in combination with an analysis of the shape of the mapped distribution. The experiments suggest further that it is advantageous to collect sensor readings along a predefined path, which roughly covers the available space and is traversed equally often from opposite directions. Reasonable results can be also achieved if the driven path satisfies these requirements approximately on average as in the Braitenberg-type experiments. In order to obtain a short stabilisation period (as with the sweeping movement where the gas sensors pass *near* by possible source locations from opposite directions very quickly) and high accuracy of the concentration maps (as with the spiral movement where each part of the trajectory was passed by equally often from opposite directions), a path that combines the mentioned properties would be a recommended candidate. A possible extension of the sweeping movement introduced in Section 5.7.2, which satisfies these requirements, is shown in Fig. 5.16.

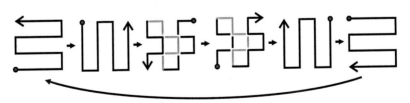

Figure 5.16: *A possible extension of the sweeping movement that satisfies the requirements for a short stabilisation period and high accuracy of the concentration maps.*

5.10   Ground Truth

In the course of the analysis in the last section it was mentioned that different situations have to be distinguished with regard to the observed gas distribution. It was assumed throughout this discussion that stabilised concentration maps represent the true average distribution, possibly with a small distortion due to the memory effect of the sensors. This assumption is difficult to prove because no direct way to verify the observed concentration field was available for the mapping experiments. One option would have been to use a dense array of fixedly installed metal oxide sensors. However, by this means it would not be possible to measure the gas distribution at the same height as with the mobile sensors carried by the robot. Furthermore, the metal oxide sensors would cause a heavy disturbance of the gas distribution due to the convective flow created by the heaters built into these sensors [ITYM03]. Another alternative way to measure a gas distribution is provided by image based methods. Here, a visible tracer gas is added and its concentration is recorded with a camera (see [NIM99], for example). For this purpose, an aerosol (smoke or mist) is typically used. While it is possible to determine the overall flow situation with this technique, it is not guaranteed that the tracer is distributed in the same way as the analyte gas. At best, tracer gas and analyte gas should therefore be identical. Apart from the problem of finding a gas that is visible and also detectable by the metal oxide sensors, the obtained two dimensional image gives an integrated view on the gas distribution that makes it difficult to separate the layer that is measured by the robot. Other possible solutions such as using GC/MS devices (gas chromatograph/-mass spectrometer) were out of question because of the high costs. Moreover, the same problem to measure the concentration in a plane at the same level as with the mobile sensors applies also to this technique.

There are, however, several indications of the consistency of the maps produced with the true gas distribution. This applies firstly to the applicability of concentration mapping for gas source localisation. The location of a gas source can be considered as some sort of ground truth about the real concentration field because it partly determines the gas distribution. In the previous section, the localisation performance was evaluated with respect to the known position of the source. It was demonstrated that the cell with the maximum value in the concentration gridmaps often provides a good estimate of the location of the gas source. This suitability for gas source localisation is an indication that the obtained representation corresponds to the real physical gas distribution.

Further indications are given by the self-consistency of the obtained concentration maps. This appears, for example, in the similarity of concentration maps that are created from different parts of the same trial. Assuming an approximately constant time-averaged structure of the gas distribution, these maps are expected to be similar. In fact, this similarity could be observed in the mapping experiments. As an example, Fig. 5.17 shows a comparison of the concentration maps obtained

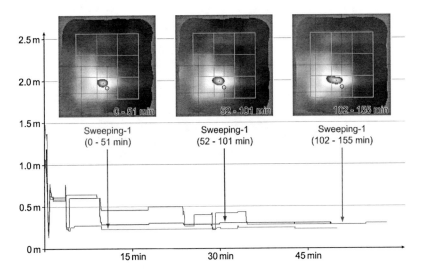

Figure 5.17: *Comparison of concentration gridmaps created from the data of different parts of the same trial "Sweeping-1". The graph shows the distance between the maximal concentration cell and the centre of the gas source. Each part corresponds to two completed cycles as defined in Section 5.7.2. The concentration maps shown in the upper part are computed from the data of one of these parts. The concentration map that results from using all the data of the experiment is shown in Figure 5.10.*

in the sweeping experiment "Sweeping-1", where the data were divided into three parts each corresponding to two completed cycles as defined in Section 5.7.2. The upper part of the figure displays the concentration maps computed from all the data of one part and the graph in the lower part shows the distance between the maximal cell and the centre of the gas source. Both views demonstrate an apparent similarity of maps created from different parts of the data, indicating both the consistency of the concentration maps with the sampled gas distribution and the fact that this gas distribution exhibits constant structures.

This type of self-consistency could also be observed with data from different trials that were carried out in quick succession and using a different data acquisition strategy. An impressive example is shown in Fig. 5.18. As in Fig. 5.17, a graph of the distance between the maximal cell and the centre of the source is depicted as well as an illustration of the concentration maps computed from all the data of the corresponding trial. The gas sensor readings were collected in two separate reactive gas source tracing experiments (with uncrossed and crossed connections),

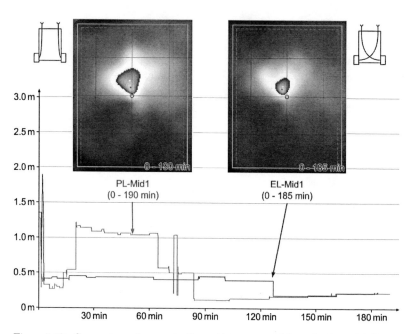

Figure 5.18: *Comparison of concentration gridmaps created from the data of different reactive gas source tracing trials ("PL-Mid1" and "EL-Mid1"), both carried out at the same day. The graph shows the distance between the maximal concentration cell and the centre of the gas source. The concentration maps in the upper part were created from all the data of the corresponding trial.*

which were carried out at the same day. Although the second trial was started more than 6 hours after the first one and the room was ventilated between the experiments, a clearly similar concentration map was obtained from the two data sets. Due to the different tracing strategies applied, the distance between the cell with the maximum concentration value exhibits a different course. However, the distance between the maximal cells found at the end of each trial was less than 10 centimetres. Similar concentration maps were generally observed in experiments with the same position of the gas source carried out on consecutive days. Apart from Fig. 5.18, such trials can also be seen in the experiments "RectSpiral-1" and "RectSpiral-2" (see Fig. D.1) or "PL-LR1" and "PL-LR2" (see Fig. D.8). A clear difference was also not discernible if separate concentration maps were created from the same trial using data from different sensors.

5.11   Online Mapping

The mapping algorithm introduced in Section 5.3 needs to be executed offline because new measurements are first normalised to the range of readings $[R_{min}, R_{max}]$ obtained in the whole experiment before they are included into a concentration map (see Eq. 5.2). Due to the linear dependency of the concentration values on the normalised readings (see Eq. 5.7 and Eq. 5.6), however, an iterative (or online) version of the mapping algorithm can be given, which yields the same result as the offline formulation considering only the time steps up to the current one.

In the following, $\tilde{r}_{t_1}^{(t_2)}$ denotes the relative reading at time t_1 that is normalised to the range of values $[R_{min,t_2}, R_{max,t_2}]$ obtained up to the time t_2. If this range changes at some time step t^*, the previously computed values $(t < t^*)$ have to be rescaled as

$$\tilde{r}_t^{(t^*)} = \frac{(R_{max,t^*-1} - R_{min,t^*-1})\tilde{r}_t^{(t^*-1)} + R_{min,t^*-1} - R_{min,t^*}}{R_{max,t^*} - R_{min,t^*}}. \qquad (5.10)$$

If a scaling is performed in this way each time the range of values changes, it is guaranteed that the computed relative concentration value always refers to the current range of sensor readings. Therefore the (online) result $\tilde{c}_t^{(i,j)}$ equals the one obtained using offline mapping up to the current time step. Hence, in order to formulate an iterative (online) mapping algorithm, Eq. 5.7 has to be replaced by the following steps. If either the lower or the upper limit of the normalisation range changes at time t, the concentration value computed for a particular sensor up to time $t - 1$ has to be rescaled as

$$\tilde{c}_{t-1}^{\prime(i,j)} = \frac{(R_{max,t-1} - R_{min,t-1})\tilde{c}_{t-1}^{(i,j)} + R_{min,t-1} - R_{min,t}}{R_{max,t} - R_{min,t}}. \qquad (5.11)$$

Using the total sum of the weights $W_{t-1}^{(i,j)}$ up to time $t - 1$, the weighting $w_t^{(i,j)}$ and the reading $\tilde{r}_t$ normalised to the current range $[R_{min,t}, R_{max,t}]$, the concentration value concerning the cell (i, j) can then be calculated as

$$\tilde{c}_t^{(i,j)} = \frac{W_{t-1}^{(i,j)}\tilde{c}_{t-1}^{\prime(i,j)} + w_t^{(i,j)}\tilde{r}_t}{W_{t-1}^{(i,j)} + w_t^{(i,j)}}. \qquad (5.12)$$

If the range of values is not changed by a new measurement, the rescaling in Eq. 5.11 is not necessary because of the equality $\tilde{c}_t^{\prime(i,j)} = \tilde{c}_t^{(i,j)}$ in this case.

It should be noted that it is necessary to maintain a separate grid of concentration values for each sensor in the online version of the mapping algorithm. In the offline version it is sufficient to store the overall concentration value obtained after each time step because different sensor channels never have to be rescaled. If, by contrast, rescaling is necessary in the online version of the algorithm, the influence of the particular sensor on the overall concentration value is reduced (because the range can only be extended). Thus, a separate value $^{n}\tilde{c}_t^{(i,j)}$ has to be stored for each sensor n and the concentration gridmap is obtained by averaging after each time step as

$$\tilde{c}_t^{(i,j)} = \frac{1}{N} \sum_{n=1}^{N} {}^{n}\tilde{c}_t^{(i,j)}, \tag{5.13}$$

with N being the total number of gas sensors.

5.12 Conclusions

This chapter presents a new technique for modelling gas distributions by constructing concentration gridmaps with a mobile robot. The introduced algorithm stores belief about the average relative concentration of a detected gas in a gridmap structure. An offline version is presented in Section 5.3 while an iterative formulation for online use is given in Section 5.11.

The concentration mapping algorithm overcomes the problem of little overlap between single measurements by using a spatial convolution of the sensor readings with a Gaussian weighting function. Thus, the decreasing likelihood is modelled that a given measurement represents the average concentration with respect to the distance from the point of measurement. The sum of the weighting functions represents the information content of a series of gas sensor measurements, especially addressing the spatial relation of single measurements and the low pass filter effect due to the long recovery time of metal oxide sensors (see Section 5.3.1). In addition to an implicit integration of succesive readings due to this "memory effect", the sensor characteristics cause a limited distortion of the resulting map, which is discussed in Section 5.4. Compared to the real distribution, the deformation appears in the mapped values as asymmetrically blurred edges and a slightly shifted centre of the area of maximum concentration. Apart from this distortion, the concentration gridmaps show time-invariant structures of a concentration field.

In order to increase spatial accuracy of the gas concentration maps, the path of a sensor should pass particular points along its trajectory from different directions. Furthermore, the path should roughly cover the inspected area, though uniform exploration is not necessary. Different data acquisition strategies are discussed in Section 5.7. The issue of parameter selection is discussed in Section 5.6. The width of the weighting function was found to be the only critical parameter. It has to be set according to the driven path by balancing the need for sufficient overlap of the measurements and the desire to resolve small structures in the gas distribution.

Finally, the new technique was evaluated using four different data acquisition strategies. Based on a total of almost 70 hours of mapping experiments the results were analysed in terms of the time needed to achieve a stable representation (Section 5.8) and the suitability for gas source localisation (Section 5.9). The gridmap representation was found to stabilise more quickly with reactive control strategies as defined in Section 5.7.3, which increase the time the robot spends in regions of high concentration. However, the structures in the gridmaps changed more often (and it took considerably longer to reach the final state of the map) compared to the experiments where a predefined path was driven because the robot could temporarily get stuck in local concentration maxima. Consequently, a predefined path (if applicable) is suggested by the mapping experiments as the superior data acquisition strategy.

By comparison with the centre of the gas source, it was demonstrated that the location of the average maximum concentration can be often used to estimate the position of a source. Because a gas source in a real world environment does not necessarily correspond to the highest concentration, the suitability of this method for gas source localisation depends on the actual airflow situation in the inspected environment. Good localisation performance therefore requires also an analysis of the shape of the mapped distribution to distinguish different airflow scenarios.

The discussion of the mapping experiments is based on the assumption that the concentration maps represent the true average distribution. Because it was not possible to verify this assumption directly by measuring the ground truth concentration with an alternative method, Section 5.10 addresses several indications for the consistency of the produced maps.

At present, only time-constant structures in the gas distribution were modelled by using temporal averaging. It would also be possible to model changing gas distributions by aging the measurements instead of averaging, so that older measurements gradually lose their weight.

Other possible developments would include experimental comparisons of different exploration strategies for map building. Strategies based on the state of the map, e.g., by moving towards areas of high uncertainty, could also be considered. Future work could also include development of an actual source finding strategy that includes a mechanism to analyse the shape of the mapped distribution.

Chapter 6

Applicability of Purely Reactive Strategies for Gas Source Tracing

"... the air doesn't smell so foul down here. If in doubt, Meriadoc, always follow your nose." (Gandalf. Lord of the Rings I, The Movie)

6.1 Reactive Localisation Strategies

In search of a light source in an unknown dark environment, it is a reasonable idea to head for the direction where the brightest glow can be sensed. Similarily, localisation behaviours in biology often involve reactive components that determine control commands only from the instantaneously sensed properties of the stimulus field [FG40]. Such behaviours do not require long-term memory because no explicit representation of the environment is needed. Consequently, biological evolution tends to favour these minimalistic solutions if they are sufficient to solve the given task. The minimal capabilities needed for reactive localisation are

- a mechanism to extract directional information from the sensed signal (tracing), and

- the ability to determine whether the source has been found (declaration).

This and the following chapter are mainly concerned with the former issue. They address the question to what extent reactive strategies are sufficient for gas source tracing in an uncontrolled indoor environment. This problem was investigated by means of different mobile robots governed towards a single immobile gas source applying different tracing strategies. In the experiments presented, the gas source was realised by a container filled with liquid ethanol. In order to prevent the robot from hitting the container, the gas source was also considered as an obstacle to the moving robot. The known position of this special obstacle was utilised for evaluation

103

purposes, but during the experiments the robot used no other information than the gas sensor measurements and an obstacle avoidance mechanism.

The following two chapters are concerned with gas source tracing behaviours that rely on reactive decisions. Strategies that implement different gradient-following behaviours are addressed in this chapter. A biologically inspired gas source tracing strategy is presented in Chapter 7. It is based on the behaviour of the silkworm moth *Bombyx mori*, and consists of a fixed motion pattern that realises a local search, and a mechanism that (re-)starts this motion pattern if an increased gas concentration is sensed. This behaviour requires a minimal memory capacity in order to execute the motion pattern, but it is still based on reactive decisions and does not involve a sophisticated representation of the environment.

All the reactive gas source tracing strategies considered were implemented on a robot and tested repeatedly in an uncontrolled indoor environment. The implementation details are discussed below and a detailed statistical evaluation of the results is given in order to analyse the performance of each strategy.

6.2 Gas Source Tracing by Gradient-Following

Unlike visual or auditory stimuli, chemical stimuli are not inherently directional. In order to achieve spatial chemo-orientation, an animal therefore has to determine a concentration gradient from comparing either successive stimuli (klinotaxis[1]) or simultaneously sensed intensities from two or more receptors (tropotaxis). As pointed out by Grasso in [HM01] the traditional framework for understanding animal orientation by Fraenkel and Gunn [FG40] is limited because it assumes a uniform gradient of the distribution. As detailed in Section 2.2, this assumption is usually not fulfilled for gas distributions in uncontrolled indoor environments. This is mainly caused by the minor influence of diffusion compared to turbulence. A real gas distribution therefore reveals many discontinous patches of local eddies, thus creating local concentration maxima that can mislead a gradient following strategy. Moreover, the absolute maximum of the instantaneous gas distribution is usually *not* situated at the location of the gas source if this source has been active for some time [LD03e]. It is generally not possible to determine the global maximum of a fluctuating distribution with many peaks by gradient-following with absolute certainty. But even if such a method was available, it would not be sufficient to declare whether the global concentration maximum corresponds to a gas source or not.

Considering the chaotic nature of turbulent gas transport, the applicability of reactive gas source tracing techniques that follow the instantaneously measured spatial gradient seems arguable. Due to the smooth course of the averaged distribution,

---

[1]The term taxis is often used to describe the movement of animals towards or away from a stimulus. It refers originally to an active body orientation with respect to a stimulus field [Küh19; FM00].

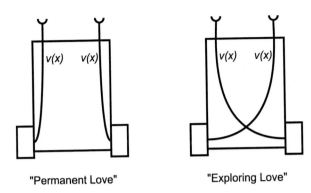

"Permanent Love" "Exploring Love"

Figure 6.1: *Schematic view of Braitenberg vehicles with a direct inhibitory sensor motor coupling. Left: uncrossed connections ("permanent love"). Right: crossed connections ("exploring love").*

however, following the gradient towards the stronger signal should be capable of driving a robot toward a gas source *on average*. The question to what degree this is true is the starting point of this investigation. In contrast to previous works [SLV93], the applicability of gas source tracing strategies based on gradient-following was tested repeatedly in a well-defined environment enabling an elaborate statistical evaluation of the results.

6.3 Braitenberg Vehicles

Two different tropotaxis behaviours were considered in this thesis that use a direct sensor-motor coupling. This kind of steering architecture is known as a Braitenberg vehicle (see Fig. 6.1) due to the influential thought experiments of Valentino Braitenberg [Bra84]. In his book Braitenberg explains which kind of behaviour should result for these vehicles (denominated as type 2, 3 and 4) using different classes of intermediate transfer functions $v(x)$ and assuming a uniform gradient. For all the experiments presented in this chapter, inhibitory connections that apply a monotonic, linear transfer function were used. Thus, maximum wheel speed results if the sensed concentration is low, which in turn implements a simple sort of exploration behaviour. On the other hand the robot is slowed down by high concentrations of the analyte.

With uncrossed connections and a monotonic, inhibitory transfer function, the wheel on the side that is stimulated more is driven slower and therefore the robot turns to this side. This behaviour was called "permanent love" by Braitenberg because it is supposed to move the vehicle to a source of stimulation and stay near this source in theory. Note that "high concentration" or "stimulation" in this context always means "high sensor values" and that these values do not reflect the actual concentration directly due to the non-zero response and the strong memory effect caused by the long decay time of the sensors (see Sections 2.1 and 3.3).

With crossed connections and an inhibitory, monotonic transfer function the robot is also slowed down by increased sensor responses but will in contrast turn away from them. Such a vehicle tends to stay at locations near to a maximum of stimulation, too, but continues to wander if another maximum comes into focus. Accordingly, Braitenberg called this kind of behaviour "exploring love". Again this description applies to a system with ideal sensors that moves guided by a smooth distribution peaked just at the actual location of any gas sources.

6.3.1   Sensor Preprocessing

The sensor-motor wiring realises a transfer function $v(x)$ that determines the speed of the connected wheel from the sensed quantity x. This quantity is supposed to represent the concentration of the analyte at the location of the corresponding sensor. In order to determine the relationship between the raw sensor readings and the corresponding absolute concentration values, calibration of each sensor is necessary because the response characteristics of individual sensors differ considerably. Note that such a calibration function applies to the steady-state case and thus cannot accomodate the non-zero response and decay time of the sensors. Calibrating gas sensors, however, is a costly operation because it has to be performed under well-defined laboratory conditions. Moreover, the whole process would have to be repeated frequently in order to compensate for the long term drift of metal oxide gas sensors. Finally, the environmental conditions, including temperature, humidity and the local airflow would have to be precisely measured to determine the absolute concentration with a robot.

For these reasons, standard calibration methods [BT95] would be not suitable for the given task. Instead, the value x is calculated by normalising the raw sensor readings R_i to lie within the range [0,1]. For this purpose, the minimum and maximum values $R_{min,i}$ and $R_{max,i}$ are constantly updated for each sensor i and used to calculate the normalised response x_i as

$$x_i^{(t)} = \frac{R_i^{(t)} - R_{min,i}^{(t)}}{R_{max,i}^{(t)} - R_{min,i}^{(t)}}. \tag{6.1}$$

Then, the normalised response values belonging to one side of the robot are combined by averaging as

$$x_L^{(t)} = \sum_{i=1}^{N_L} x_i^{(t)}/N_L, \qquad x_R^{(t)} = \sum_{i=1}^{N_R} x_i^{(t)}/N_R, \tag{6.2}$$

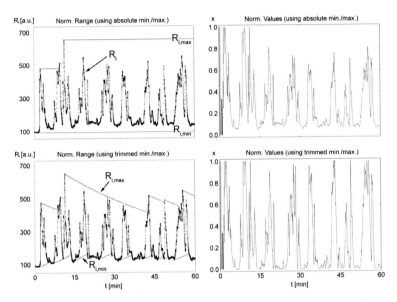

Figure 6.2: *Comparison of normalisation techniques. Upper row: normalisation without trimming. Lower row: normalisation with trimming. Left: raw sensor readings and the minimum/maximum values used to calculate the normalised value x. Right: normalised value x.*

where N_L and N_R are the number of sensors contained in the sensor array on the left and right side respectively. For the experiments presented in this chapter, $N_L = N_R = 3$ sensors were used.

Because it eliminates the need for external intervention, such a normalisation technique provides a practical solution for possible applications, which is able to compensate for the sensitivity mismatch of individual sensors as well as for seasonal and environmental drifts. Rather than the absolute concentration of the analyte, the normalised value x represents the relative concentration with respect to the current situation[2].

While it is possible to compensate for different conditions in individual runs, the normalisation technique described so far is unable to adapt to varying conditions during a single trial. In response to a changing temperature, for example, the normalisation range gets wider and might not cover the actual range of values with

---

[2]Note that the assumption of a linear resistance-concentration dependence is made here.

time. This causes changes in response to be less pronounced in x. Note that the same effect occurs naturally if the average concentration is rising due to the presence of an active gas source. To avoid this problem, the normalisation range is dynamically trimmed each Δt^{trim} seconds by increasing the minimum and decreasing the maximum value in Eq. 6.1 by a fixed fraction of the normalisation range as

$$R^{(t)}_{min,i} = \tilde{R}^{(t)}_{min,i} + \Delta^{trim}_{min}(\tilde{R}^{(t)}_{max,i} - \tilde{R}^{(t)}_{min,i}), \tag{6.3}$$

$$R^{(t)}_{max,i} = \tilde{R}^{(t)}_{max,i} - \Delta^{trim}_{max}(\tilde{R}^{(t)}_{max,i} - \tilde{R}^{(t)}_{min,i}). \tag{6.4}$$

Here, $\tilde{R}^{(t)}_{min,i}$ and $\tilde{R}^{(t)}_{max,i}$ refer to the minimum and maximum value at time t before trimming.

A comparison of both normalisation techniques is depicted in Fig. 6.2 on the basis of gas sensor data collected at the beginning of an experiment, where the robot was steered as a Braitenberg vehicle with uncrossed connections. The normalisation range and the raw sensor values R of a gas sensor of type TGS 2600 are shown in the panels on the left side, while the resulting values of x are given on the right side. It can be seen that it takes some minutes until the normalisation range gets initially widened by the concentration peaks at the beginning. In the following minutes, the normalised value approximately reflects the relative concentration with respect to the range of concentration values occuring at that time. However, if the normalisation range is not trimmed (see the upper row in Fig. 6.2), the amplitude of the normalised value does not cover the full interval $[0,1]$ after some time. In this example, this effect was caused by the outstanding peak that occured approximately 11 minutes after the experiment started and by the rising level of the minimum gas concentration sensed. As can be seen in the lower row of Fig. 6.2, the normalisation range adapts to this change if trimming is used. Consequently, the normalised value x represents the relative concentration with respect to the concentration extremes that occur around that time. In the all experiments described in this chapter, the trimming parameters

$$\Delta^{trim}_{min} = \Delta^{trim}_{max} = 1\% \text{ and } \Delta t^{trim} = 30 \ s$$

were used, which were found to be suitable in an initial test sequence.

6.3.2   Transfer Function

In all the experiments presented below, a linear inhibitory transfer function was used as

$$v(x) = K_v(1 - x), \tag{6.5}$$

with a velocity gain K_v of either 5 cm/s or 3 cm/s.

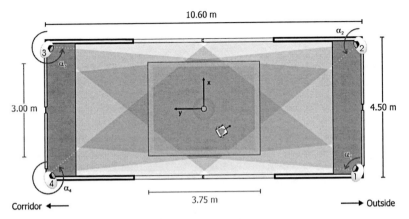

Figure 6.3: *Testbed area for gas source tracing experiments. The figure shows a floor plan of the laboratory room (including doors and windows) and the outline of the region in which the experiments were performed. Additionally, the repellent potential field is indicated at the edge of the testbed area. Also plotted are the fields of view for each of the four cameras that were used to track the robot's position, shaded according to the number of cameras which can sense a particular region.*

6.4 Experimental Setup

The experiments to test the Braitenberg-type gas source tracing strategies were performed in a rectangular laboratory room at Örebro University (size 10.6×4.5 m^2). A floor plan of this room is shown in Fig. 6.3. The air conditioning system in the room was deactivated in order to eliminate the possibility of a dominant constant airflow. All the runs were conducted in the same room and the environmental conditions were kept as constant as possible.

The test platform was chosen to be a Koala mobile robot equipped with the Mark III mobile nose. This comparatively small robot is described in Section 3.2.1, while the Mark III mobile nose is detailed in Section 3.2.2. Fig. 3.1 shows the robot, carrying the mobile nose and a cardboard hat that is used by the absolute positioning system W-CAPS in order to track the robot's position (see Appendix B for details of the positioning system).

To emulate a typical task for an inspection robot, the gas source was chosen to imitate a leaking tank. This was realised by placing a paper cup filled with ethanol on a support in a bowl with a perimeter of 12 cm (see Fig. 3.1). The ethanol dripped through a hole in the cup into the bowl at a rate of approximately 50 ml/h. Ethanol was used because it is non-toxic and easily detectable by the tin oxide sensors.

6.4.1   Testbed Scenario

Both Braitenberg-type strategies considered, which can be also described as "exploration and hillclimbing" (Fig. 6.1, left) and "exploration and concentration peak avoidance" (Fig. 6.1, right), were tested repeatedly with the following scenario. A 3.75×3 m^2 field was defined by establishing virtual walls. These boundaries were realised by assigning an artificial potential field [Kha85] that effects a repellent pseudo-force, which increases linearly with the penetration depth and starts to be effective at a distance of 20 cm. The testbed area was defined in the central region of the room where precise and reliable position information is available. Both the virtual walls and the area, where the repellent pseudo-force is active, are shown in Fig. 6.3.

Now the robot could move freely within this virtual field while being constantly tracked by the positioning system W-CAPS (see Appendix B for details). Next, a gas source was placed at a known position inside the field. A successful trial is counted whenever the robot enters the clearance area around the gas source. Similarly, a successful trial is counted in reference tests where no gas source is used, if the robot enters the same clearance area. A series of experiments with different tracing strategies were performed as follows:

- set the robot to a random starting position inside the virtual field (with a clearance of at least 100 cm to the centre of the source),

- rotate the robot to a random initial heading,

- start to move the robot controlled by the particular strategy to be tested,

- count a successful try and restart if the robot enters the obstacle clearance area around the gas source.

These steps were repeated for a fixed amount of time while the position and the sensor readings were constantly logged for evaluation purposes.

6.5   Results

The following sections present a discussion of the behaviour of gas-sensitive Braitenberg vehicles with uncrossed ("permanent love") and crossed inhibitory connections ("exploring love"). Typical trajectories are shown in Figures 6.4 – 6.7. Here, the path of the robot's centre is indicated by small circles, while the position of the front corners is plotted using small dots. The starting position and the initial heading of the corresponding trial are marked by an arrow, which originates from the starting position. Also shown are the virtual repellent walls (broken line) that enclose the area where the repellent force increases with the penetration depth of the robot. Finally, the clearance area of the gas source is shown by two circles. A trial was stopped if one of the front corners of the robot entered the inner circle. The outer circle was derived by obstacle growing, assuming a circular shaped robot. The radius was chosen to be the minimum distance between the centre of the robot and the source at which a collision in the above defined sense can occur. Note that because the robot actually has a rectangular profile, the outer circle provides just an approximate notion of the obstacle's boundary with respect to the centre of the robot. The trajectories of all the trials with a Braitenberg-type strategy are shown in Appendix E.

Interpretation Using Concentration Gridmaps

In order to explain the resulting trajectories, it would be helpful to know the transient gas distribution at each time step. As discussed in Section 5.10, however, it was not feasible in this work and is generally very difficult to measure the entire concentration field at the same time and at the same height as with the mobile sensors carried by the robot without considerable disturbance of the gas distribution. While the transient distribution cannot be measured with the gas sensors carried by the robot, its time-invariant structure can be determined using the technique to create gridmaps of the relative gas concentration, which was introduced in Chapter 5.

In addition to the typical trajectories shown on the left side of Figs. 6.4 – 6.7, the same trajectory is shown on the right side on top of the corresponding gas concentration gridmap. These gridmaps were created offline with a cell size of 2.5×2.5 cm^2 from the gas sensor readings recorded during all the trials belonging to one experiment. Limited by the available battery capacity, complete experiments lasted approximately 3 hours. To illustrate these gridmaps, the relative concentration values are indicated by shadings of grey (dark $\rightarrow$ low, light $\rightarrow$ high). In order to represent plume-like structures, values higher than 90% of the maximum are plotted with a second range of dark-to-light shadings {of red}.

6.5.1 Gas Source Tracing by Exploration and Hillclimbing

In the case of uncrossed connections ("permanent love"), the robot could frequently reach the source in a strikingly straightforward way, as in the example trial shown in Fig. 6.4. In this experiment the source was placed in the middle of the virtual field. The robot was almost unaffected by the gas source until it bumped into the virtual wall for the fifth time. As can be seen on the right side of Fig. 6.4, the behaviour before the fifth collision is consistent with the concentration gridmap, as the robot remained outside the region of high concentration during that time. Then, the robot entered the region of highest concentration (the plume), slowed down, turned to the left side (where it had entered the plume first), and finally hit the clearance area around the gas source.

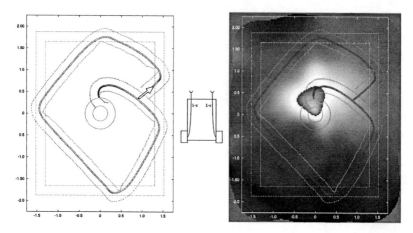

Figure 6.4: *Example of the path of a "permanent love" Braitenberg vehicle (uncrossed (1-x)-connections) that corresponds to the average gas distribution. The gas source is located in the middle of the testbed area.*

While the previously mentioned trial could be explained in terms of the average distribution, this is not always possible. The trajectory shown in Fig. 6.5, for example, is likely to be caused by transient concentration fluctuations. This trial was performed directly after the one shown in Fig. 6.4. Here, the robot passed the gas source immediately after the trial was started at a very low distance. However, hardly any reaction was obtained, neither concerning the translational nor the rotational speed. The path during the remainder of the trial appears consistent with the gas concentration map. The robot orbited around the gas source and moved along roughly straight lines outside the region of high concentration. Ultimately the robot entered the plume (with the right side first), turned back and slowed down, and finally moved slowly towards the source.

Source	Strategy	$K_v[\frac{cm}{s}]$	l_{tot} [m]	$\bar{l}$ [m]	$\bar{d}$ [cm]	$\bar{v}/K_v$
Middle	Ref (1-x)	5	319.0	9.67 ± 7.66	136.7 ± 44.9	95.9%
	PL, 1-x	5	1044.0	8.49 ± 7.93	127.4 ± 45.7	73.5%

Table 6.1: *Statistics of the gas source tracing experiments with a Braitenberg vehicle and the gas source placed in the middle of the testbed area. Comparison of the "permanent love" behaviour (PL) with the reference strategy (Ref).*

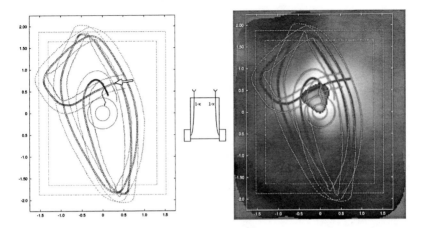

Figure 6.5: *Example of the path of a "permanent love" Braitenberg vehicle (un-crosssed (1-x)-connections) that does not correspond to the average gas distribution. The gas source is located in the middle of the testbed area.*

The statistical evaluation of the experiments, with the gas source placed in the middle of the testbed area, is summarised in Table 6.1. Here and in the following Tables 6.2 and 6.3, the leftmost column specifies the position of the gas source while the second column references the applied strategy, denoted as Ref (reference experiment), PL (uncrossed connections, "permanent love") or EL (crossed connections, "exploring love"). The remaining columns itemise (from left to right): the speed gain K_v (see Eq. 6.5), the total path length l_{tot}, the average path length of successful trials $\bar{l}$, the average distance to the source $\bar{d}$, and the average velocity $\bar{v}$ in units of the speed gain. To validate the results, reference tests were performed without a gas source. Because the sensor readings were not considered during these tests, the robot moved basically like a ball on a billiard table. As in all other experiments, a successful trial was counted when the robot entered the area assigned to be the source.

With the source placed in the middle of the inspected area, the average path length of a vehicle with uncrossed (1-x)-connections was 8.49 m compared to a reference value of 9.67 m (the median was 6.57 m compared to 8.31 m). Applying the Wilcoxon two sample test, no statistically significant difference could be determined in this case ($p_{H_0} = 0.2685$)[3]. (Because the path lengths are not normally distributed, a statistical test was performed that does not assume a specific distribution.) Thus,

---

[3]The Wilcoxon two sample test evaluates the hypothesis H_0 that the populations from which the samples are taken have identical median values.

Source	Strategy	$K_v[\frac{cm}{s}]$	l_{tot} [m]	$\bar{l}$ [m]	$\bar{d}$ [cm]	$\bar{v}/K_v$
Corner	Ref (1-x)	5	1554.9	20.46 ± 19.38	223.5 ± 78.7	97.6%
	PL, 1-x	5	733.3	14.10 ± 14.56	217.8 ± 90.1	81.2%
	PL, 1-x	3	517.8	9.42 ± 6.84	183.8 ± 84.9	71.3%
	PL, 1-x	5/3	1251.1	11.69 ± 11.22	199.3 ± 88.2	75.5%

Table 6.2: *Statistics of the gas source tracing experiments with a Braitenberg vehicle and the gas source placed in a corner of the testbed area. Comparison of the "permanent love" behaviour (PL) with the reference strategy (Ref). If different speed gains K_v were considered for evaluation, both of them are given separated by a slash.*

a significant improvement in the gas source tracing performance in terms of the average path length cannot be asserted. However, it is important to note that the tested reference strategy, which approximately implements specular reflection at the walls, does not provide uniform coverage over a designated area [Gag93]. Instead, the central area is covered excessively at the expense of the periphery [McN87]. Hence, the fact that the Braitenberg vehicle does not outperform the reference strategy might be caused by the inherent tendency of the reference strategy to "find" a gas source in the middle of the testbed area.

Additional tests were therefore performed with the source placed at a less prominent location near to one corner of the field (15 cm away from the beginning of the repellent wall potential along both the x- and y-axis). Two typical runs of these experiments are shown in Fig. 6.6. The trajectory in the upper part corresponds nicely with the concentration gridmap. After the robot got "caught" by the plume-like structure, it drove very slowly towards the source following an almost straight line. Similarly, the path of the robot that is shown in the lower part of Fig. 6.6 complies with the corresponding concentration gridmap if only the behaviour in the direct vicinity of the gas source is considered. The stretched right hand bend, however, that leads the robot towards the right corner cannot be explained by this means.

The results of the experiments with the gas source placed in a corner of the testbed area are summarised in Table 6.2. For each corner a total of approximately 3 hours of trials was performed both with and without a source (the duration of all trials is given in Appendix E). The average path length needed to reach an active gas source was 11.69 m, compared to 20.46 m in the reference experiments (median: 7.25 m compared to 17.09 m). In contrast to the situation where the gas source was placed in the middle of the testbed area, the Wilcoxon two samples test reveals a highly significant improvement in tracing performance in terms of the average path length ($p_{H_0} < 10^{-4}$).

This result clearly shows that following the instantaneous concentration gradient can improve the gas source tracing performance compared to search strategies that

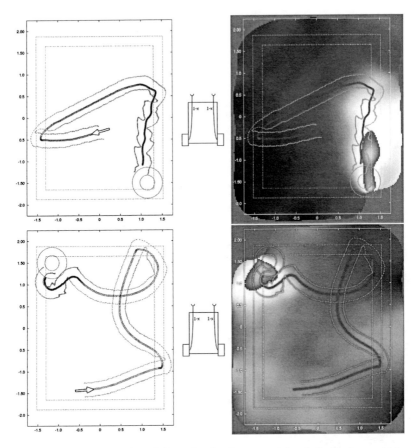

Figure 6.6: *Two examples of the path of a "permanent love" Braitenberg vehicle (uncrosssed (1-x)-connections) with the gas source in a corner of the testbed area.*

do not consider cues from the sensed gas distribution. Note that a better performance in this context means that the robot reaches the source along a shorter path and not necessarily in shorter time. This follows from the dependency of the gain in "path length performance" on the applied speed gain K_v, which is a consequence of the non-zero response time of the metal oxide gas sensors (discussed in Sections 2.1 and 3.3). Although this conclusion cannot be drawn with high statistical certainty from the experiments with $K_v = 5$ cm/s and $K_v = 3$ cm/s ($p_{H_0} = 0.9727$), it is

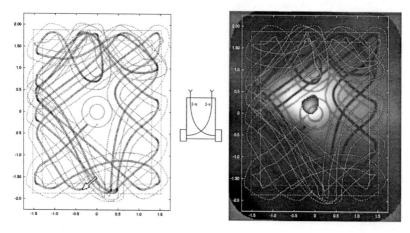

Figure 6.7: *Example of the path of a Braitenberg-Vehicle with crosssed (1-x)-connections ("exploring love") with the gas source in the middle of the testbed area.*

apparent that a Braitenberg-type strategy, which relies on measurements that reflect the stimulus with a certain delay, should perform less well if the speed gain K_v increases. Thus, a standard search strategy will perform better in terms of time if it operates at sufficiently high speed. In fact even at the rather low speeds investigated, the reference strategy performs comparably well in this respect: the average time to reach the source was 419.3 s for the reference search, compared to 348.6 s ($K_v = 5$ cm/s) and 444.1 s ($K_v = 3$ cm/s) for the Braitenberg vehicles. Thus, the reference strategy can outperform both the "permanent love" behaviours tested in terms of the time needed to reach the gas source, if a speed gain of approximately $K_v > 6$ cm/s is applied.

6.5.2   Exploration and Concentration Peak Avoidance

With crossed connections a completely different behaviour results (see Table 6.3). Although the robot is expected to stay near the source – again assuming a smooth gradient – and thus collisions should not be unlikely, the robot managed to avoid the source most of the time (see Fig. 6.7). The difference compared to the trials with uncrossed connections is apparent and can be demonstrated with high certainty by a Wilcoxon two sample test ($p_{H_0} < 10^{-7}$).

Though this strategy is obviously not suitable for driving the robot towards a gas source as quickly as possible, it offers an interesting solution to the full gas source

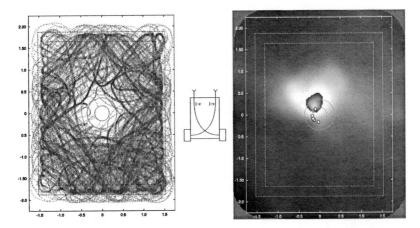

Figure 6.8: *Left side: path of a Braitenberg-Vehicle with crosssed (1-x)-connections ("exploring love") during a complete experiment, including 3 hours and 5 single trials. Right side: Concentration gridmap determined from the readings collected during the same experiment. The locations where the robot entered the clearance area around the gas source are indicated by small circles.*

localisation problem, including the declaration of the source. This can be seen in Fig. 6.8, which shows the complete path of the robot during the experiment, of which one trial (out of five) is shown in Fig. 6.7. After combining the paths of all the trials in this experiment, the location of the source is indicated by the part of the picture that remains light. Notice that in contrast to the area that is covered by an obstacle (the inner circle in Fig. 6.8), the area between the concentric circles should not remain completely light in the case of pure obstacles.

Source	Strategy	$K_v [\frac{cm}{s}]$	l_{tot} [m]	$\bar{l}$ [m]	$\bar{d}$ [cm]	$\bar{v}/K_v$
	Ref (1-x)	5	319.0	9.67 ± 7.66	136.7 ± 44.9	95.9%
Middle	PL, 1-x	5	1044.0	8.49 ± 7.93	127.4 ± 45.7	73.5%
	EL, 1-x	5/3	731.8	40.66 ± 34.10	143.3 ± 41.0	76.5%

Table 6.3: *Statistics of the gas source tracing experiments with a Braitenberg vehicle and the gas source placed in the middle of the testbed area. Comparison of the "permanent love" behaviour (PL) with the "exploring love" behaviour (EL) and the reference strategy (Ref). If different speed gains K_v were considered for evaluation, both of them are given separated by a slash.*

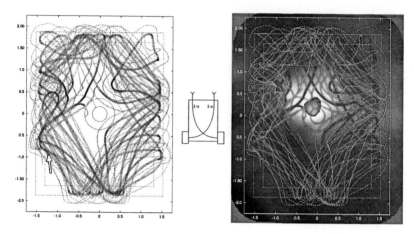

Figure 6.9: *Example of the path of a Braitenberg-Vehicle with crosssed (1-x)-connections ("exploring love") with the gas source in the middle of the testbed area.*

The reason why exploration and concentration peak avoidance might be a solution for gas source localisation can be explained as follows: with crossed inhibitory connections the robot explores the available space and evades each local concentration maximum. Because there exist many of them it is hard to find a particular maximum that belongs to the actual source by a hillclimbing strategy. On the other hand, concentration maxima occur more frequently near the gas source and thus the source's location remains comparatively light in a plot of the driven path such as Fig. 6.8.

Note that the experiments point out a feature that can be used to identify a gas source rather than providing a complete solution that is generally applicable. If the source is not detectable as an obstacle, for example, it would not be possible to avoid collisions with the currently used setup. During the experiments the robot moved occasionally into the clearance area around the gas source. As can be seen on the right side of Fig. 6.8, which shows the position where the robot entered the clearance area, this happened mostly if the robot approached the gas source from the opposite side of the plume-like structure (an example of such a collision is shown in Fig. 6.7). The trial that cannot be classified in this way is shown in Fig. 6.9. Here, the turn to avoid an increasing concentration was started too late and the robot would not have been able to avoid a collision with the gas source if the corresponding trial was not externally stopped. A collision with the gas source could corrupt the practicability of the proposed method if the robot spilled the smelling liquid or wetted its tyres, which would cause the robot to become a gas source, too.

6.6 Conclusions

Two Braitenberg-type gas source tracing strategies were investigated in this chapter. Within a total of 36.5 hours of experiments where the robot drove almost 5 kilometers, both strategies were shown to be of possible use for gas source localisation. Using uncrossed inhibitory connections it was found that the average path length the robot needs to move to the source can be decreased. The path length could be reduced by up to a factor of two compared to a strategy that explores the available area ignoring the gas sensor readings. This factor, however, depends on the considered situation and the reference search strategy used.

For real world applications this strategy has to be extended by an additional declaration mechanism to determine that the gas source has been found with high certainty. This mechanism could be added by using other sensors, which provide clues on possible sources, for example, by recognising a beaker or a puddle by vision. A source declaration method based on gas sensor readings only is discussed in Chapter 8.

With crossed connections the robot evades each local concentration maximum, including those that are closest to the source. Due to the fact that maxima occur more frequently near the gas source, the path of the robot covers the whole available area except that near to the actual location of the source.

Applying this strategy therefore offers a solution to the task of gas source declaration without using additional sensors. A further advantage is that the average distance to the gas source is increased (see Table 6.3) and that direct contact between the robot and the volatile or liquid substance to be detected can be diminished. This might be preferable in the case of an analyte that is corrosive or in other ways offensive to the robot itself. Furthermore, it helps to prevent the robot from wetting its wheels in cases where the gas source is a liquid substance that cannot be sensed as an obstacle. This is generally not desired because the robot would thereby become a gas source, too. With the setup of the Mark III mobile nose, however, it was not possible to avoid collisions with the gas source completely. Future work should therefore address possible modifications to achieve a collision-free path even if the gas source cannot be sensed as an obstacle. A suggestion would be to add a third gas sensor array in a tube that sticks out to the front of the robot in order to examine the area the robot is about to drive towards. If this third tube protrudes over the robot's front sufficiently, it should be possible to decrease the probability of traversing a puddle of the analyte with the robot, even if the robot approached the source from a down-wind direction.

A possible objection to the suggested method is time consumption. Because the actual location of a gas source is determined by excluding all other possible locations, the time needed to locate the source increases with the size of the area observed. On the other hand, the time consumption scales down with the number of robots utilised. A reduction of the required duration could be further achieved by extending the random exploration strategy, which is applied by a Braitenberg vehicle, with a strategy that implements a tendency to drive towards currently unexplored areas. And after all, there is as yet no other known method to localise a gas source that does not appear as an obstacle to the robot in an uncontrolled environment.

Chapter 7

Biomimetic Gas Source Tracing

"Remarkably, the movement tracks that many species produce when tracking odour plumes have a very similar side-to-side zigzag shape whether walking, swimming, or flying." (Mark Willis [Wil])

7.1   Learning From Animals

Probably all living creatures respond to chemical stimuli. The corresponding senses are thought to be ancient, because the fundamental structures are similar in species all the way from fish to insects to primates [Han96]. By using their sense of smell, higher animals acquire information from their environment. For many of them, olfaction is basic to the maintenance of life [WK01]. While repulsive odours are often used to avoid possibly dangerous situations, attractive odours enable animals to localise food (like sea-birds that trace the airborne plume of fish oil in order to find their prey), or a mate (like moths, where the male localises a female conspecific guided by pheromones [KL59; BBSH59]), for example. The importance of olfactory cues in the animal kingdom corresponds to impressive gas source localisation capabilities. Male moths are able to detect only a few pheromone molecules while it takes about 200 molecules to start their pheromone tracing behavior [Sch69]. They were shown to locate synthetic pheromone blends over a distance of one kilometer [PWV86].

Providing mobile robots with equivalent capabilities to localise a gas source would enable a broad range of applications, including humanitarian demining, surveillance and rescue missions. Strategies based on gradient-following, however, are prone to be misled by the non-smooth and multimodal character of gas distribution in real world environments (see Chapter 6). The turbulent nature of gas transport typically leads to a distribution that consists of comparatively isolated patches with high concentration of the analyte gas, surrounded by areas where the concentration is considerably lower [MEC92; HMG02]. Nevertheless, animals are known to be

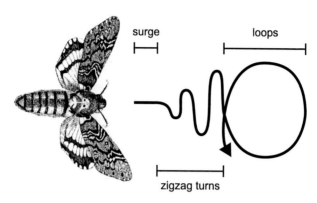

Figure 7.1: *Schematic view of the gas source tracing behaviour of the Bombyx male following a figure in [Kan96]. Note that the depicted moth icon is not intended to give a biologically correct view of the Bombyx mori species.*

very efficient in localising odour sources. Their gas source tracing behaviours have evolved over a long time to reach an optimal state concerning the particular sensor equipment of the animal.

Although biological sensors differ greatly from those used by mobile robots, the general principles applied by biological systems can be transferred to mobile robots. In this sense, a biologically inspired approach for gas source tracing is presented in this chapter. Here, an algorithm is used that is based on the strategy to search for a pheromone source applied by the male silkworm moth *Bombyx mori*. When this moth detects pheromones released by conspecific females, it uses the local wind direction as an approximation of the direction to the gas source in order to align an oriented local search for the next pheromone patch. For the implementation on a real robot, however, an indoor environment without a strong unidirectional airflow is considered, where it was not possible to measure the local wind speed accurately because of the limitation of currently available anemometers (see Section 2.8). Therefore, an adapted method was implemented that uses gas sensor readings only.

The rest of this chapter is structured as follows. First, the biological behaviour is introduced, which is applied by male *Bombyx* moths to localise a female mate as a source of pheromones (Section 7.2). After a description of the experimental setup (Section 7.4), the adapted strategy is detailed in Section 7.5. Then, a few remarks on the experimental procedure are given in Section 7.6. Finally, the results are presented in Section 7.7, including a detailed statistical evaluation in order to analyse the performance of the adapted strategy, followed by conclusions and suggestions for future work (Section 7.8).

7.2 Tracing Behaviour of *Bombyx Mori* Males

The gas source tracing behaviour of the silkworm moth *Bombyx mori* is well-investigated and suitable for adaptation on a wheeled robot, because this moth walks rather than flies in order to follow female sex-attractant pheromones [Kan98a]. The behaviour results from a combination of reactions to peripheral input and centrally organised self-generated manoeuvres [Kan98b]. It is based mainly on three mechanisms [Kan96]:

- *a trigger*: if the moth's antennae are stimulated by intermittent patches of pheromones, a fixed motion pattern is (re-)started,

- *a local search*: the motion pattern realises an oriented local search for the next pheromone patch, and

- *a mechanism to estimate the direction to the source*: the main orientation of the motion pattern that implements the local search is given by the instantaneous upwind direction, which provides an estimate of the direction to the source.

Stimulation of either antenna triggers the specific motion pattern of the *Bombyx* males. This fixed motion sequence (see Fig. 7.1) starts with a forward surge directed against the local air flow direction. Afterwards, the moth performs a "zigzag" walk, while it starts to turn to that direction where the stimulation was sensed. The turning angles and the length of the path between subsequent zigzag turns is increased with each turn [Kan96]. Finally, a looping behaviour is performed, while complete cycles with more than 360° are possible. This "programmed" motion sequence is restarted from the beginning if a new patch of pheromones is sensed. As could be shown in wind tunnel experiments by Kanzaki [Kan96], this behaviour results in a more and more straightforward path directed towards the pheromone source if the frequency of pheromone stimulation is increased.

7.3 Related Work

An implementation of the pheromone-oriented behaviour of the *Bombyx* males on a mobile robot was tested by Russell and co-workers [RBHSW03]. In contrast to this work, a considerably smaller robot with a diameter of 10 cm was used and the experiment was carried out in a smaller, strongly ventilated environment (the airflow was approximately 1.5 m/s), which enabled to determine the wind direction with an anemometer. The applied strategy was implemented in an iterative manner, as described in Section 2.7. In the experiment carried out on the top of a table tennis table, the robot could successfully trace the gas source in seven out of ten trials over a distance of 2 m. The length of the tracing path was relatively short ($\approx 130\%$ of

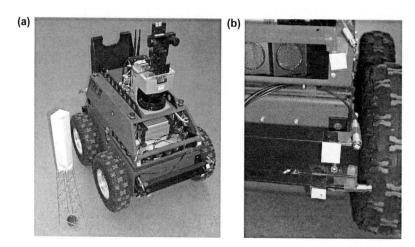

Figure 7.2: *(a) The mobile robot Arthur equipped with two sets of three gas sensors on each side at the front corner. Also shown is the gas source to the left of the robot. (b) Position and orientation of the gas sensors.*

the shortest possible path) according to the fact that the local search was usually restarted before the looping phase.

7.4   Experimental Setup

7.4.1   Robot and Gas-Sensitive System

The adapted gas source tracing strategy was implemented on the mobile robot Arthur, which is described generally in Section 3.4.1. Aside from the gas sensitive system, only the data from the laser range scanner were used to correct the position data obtained from odometry. In order to minimise self-produced disturbance of the gas distribution, the robot was supplemented with a device to deflect the air stream that is produced by the two fans at the back side. For this purpose, a dark plastic storage rack, which can be seen in 7.2 (a), was mounted above the rear fans with its opening directed upwards. Note that the robot (length = 80 cm, width = 65 cm, height without laser range scanner = 55 cm) is considerably bigger than a silkworm moth, where the antennae are separated by just a few centimeters.

The mobile nose is based on the commercial gas sensor system VOCmeter-Vario from AppliedSensor described in Section 3.4.2. For the experiments presented in

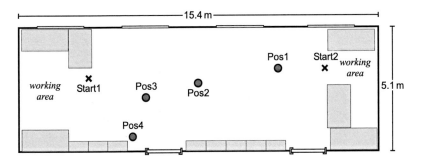

Figure 7.3: *Floor plan of the room used for the "robotic moth" experiments. Indicated are the positions where the robot was started: "Start1" at (2.7 m, 2.2 m) and "Start2" at (13.3 m, 3.35 m) with respect to the lower left corner of the room, as well as the four source positions tested: Pos1 at (11.0 m, 3.25 m), Pos2 at (7.65 m, 2.8 m), Pos3 at (5.45 m, 2.2 m) and Pos4 at (4.95 m, 0.6 m).*

this chapter, two sets of three metal oxide sensors (of type TGS 2600 from Figaro) were used. These sensors were symmetrically mounted at a height of 21 cm, 29 cm and 40 cm on each side at the front corners of the robot with a spacing of 50 cm. The sensors were aligned differently (see Fig. 7.2 (b)), in order to reduce possible effects of the direction of motion.

7.4.2 Environment and Gas Source

All experiments were performed in a 15.4 m × 5.1 m room at the University of Tübingen. A floor plan of this room is shown in Fig. 7.3. In addition, the obstacles in the room (cupboards and desks), the starting positions of the robot ("Start1", "Start2") and the position where the gas source was placed ("Pos1" – "Pos4") are indicated. With regard to real world applications, the environment was not severely modified for this investigation. The unventilated room was also used as an office where up to two persons were working during the experiments. While the windows were always kept shut, it was accepted that the persons were moving and sometimes leaving or entering the room. Although they were told to be careful, this indoor environment can be considered uncontrolled to some extent.

The gas source was chosen to be a cylindric vessel with a diameter of 40 mm and a height of 25 mm filled with ethanol, which was used because it is non-toxic and easily detectable by metal oxide sensors. In order to be recognisable by the laser range scanner, a frame made of wire with a cardboard marking mounted on top was placed above the vessel (see Fig. 7.2 (a)).

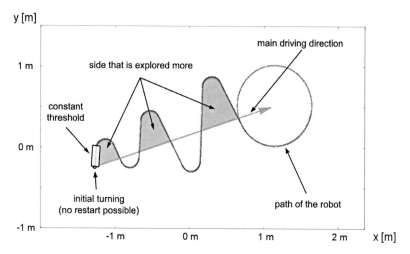

Figure 7.4: *Gas source tracing strategy adapted for a mobile robot. Fixed motion pattern that is executed by the robot in response to an increased gas concentration.*

7.5 Modified Gas Source Tracing Strategy

7.5.1 Fixed Motion Pattern

For the implementation on a real robot, the gas source tracing strategy of the moth *Bombyx mori* had to be adapted because information about the local wind direction was not available. In the algorithm that is used by the robot, those parts of the biological strategy that rely on the local air stream direction are therefore skipped. Consequently, the robot's orientation is not changed at the beginning of the fixed motion pattern and the initial forward surge is omitted. The resulting path is shown in Fig. 7.4. After being triggered by increased sensor readings, the robot starts the zigzag movement by turning approximately 65° to the side at which the higher concentration was sensed. Afterwards it performs six zigzag turns (with a length of the successive straight movements of approximately 20 cm, 30 cm, 50 cm, 70 cm, 90 cm, and 55 cm, respectively), followed by a circular motion with a radius of approximately 50 cm. Note that merely approximate values can be given because the motion pattern is executed using the odometry only.

The main direction of the zigzag motion is equal to the current heading of the robot when the fixed motion pattern is triggered. However, the robot can change its orientation if the motion sequence is restarted. Because the asymmetric path is biased towards the side where the higher sensor readings were obtained, this side is explored more (see Fig. 7.4). Consequently, the chance that the motion pattern is restarted on this side is higher. Assuming that there is a higher chance to find the gas source on the side where the stronger concentration is sensed, a robot controlled by this gas source tracing strategy should be able to move towards a gas source and to stay within close proximity to it.

7.5.2 Trigger

Sensor Preprocessing

In order to compensate for the sensitivity mismatch of individual sensors as well as for seasonal drifts, the raw sensor readings r_i were normalised to the range of [0,1] using dynamical trimming as described in Section 6.3.1. As in the case of the gas source tracing experiments using gradient-following (discussed in Section 6), the trimming parameters

$$\Delta_{min}^{trim} = \Delta_{max}^{trim} = 1\% \text{ and } \Delta t^{trim} = 30 \ s$$

were also used for the experiments presented in this chapter.

Releasing the Trigger

The trigger that (re-)starts the motion pattern operates on the normalised sensor readings. It is released if the normalised value calculated with Equations 6.1 – 6.4 exceeds a threshold that is initially set to a value χ_0.

As a consequence of the long decay time of the metal oxide gas sensors, a prolonged period of increased sensor readings results if the robot enters an area with a high gas concentration. To avoid permanent triggering due to this effect, two additional mechanisms are used: first, the motion pattern cannot be triggered during the initial turn (before the robot starts to drive along the first straight line). In addition, the threshold is set to the value that triggered the motion sequence and kept constant until the robot reaches the end of the first straight track. After the first straight movement has been completed, the threshold is then decreased at a rate of $\Delta \chi$ until it reaches the minimum value χ_0. For the experiments presented below, values of $\chi_0 = 0.2$ and $\Delta \chi = 0.01 \ s^{-1}$ were used, which were found to be suitable in an initial test sequence. By changing the trigger threshold in this way, the patchy structure of the gas distribution is reproduced in the sensor feedback, which appears blurred due to the response charactistics of the gas sensors. Both applied mechanisms are indicated in Fig. 7.4.

7.5.3   Randomized Search

As long as increased sensor readings are not obtained, the robot explores the available area by using a randomized search strategy. The robot drives along straight paths until it enters the clearance area around an obstacle. If so, a direction is randomly chosen from the set of valid options. The robot then rotates to this direction and proceeds with a straight movement.

If the robot enters the clearance area around an obstacle while performing the fixed motion pattern, the translation speed is set to zero (without changing the rotational fraction) until it is able to continue driving.

7.6   Experiments

In order to avoid a gas distribution that is dominated by the gas cloud that might result from the pouring process, the vessel was filled approximately two hours before the experiment started and covered with a plastic sheet. After the room had been ventilated by opening the windows for one hour, temperature and airflow disturbances were compensated by waiting another hour, then the sheet was removed and the experiment started.

Before uncovering the vessel, the robot was placed at one of the positions indicated by the label "Start1" and "Start2" in Fig. 7.3. Then, the robot was started, whereas a random forward direction was chosen by the exploration behaviour. The velocity was limited to 4 cm/s. In order to be able to escape U-shaped obstacles a clearance of 85 cm had to be used. Therefore, this represents the minimum distance to the gas source the robot could reach during the experiments.

7.7   Results

In the Figures 7.5 and 7.6 the path of the robot during four tracing trials is shown. In order to obtain accurate information about the path driven, a scan matching algorithm [BS03] was used for offline correction of the odometry data.

In the trial that is shown on the left side of Fig. 7.5, an active gas source was placed at "Pos1". After a short period of exploration, the trigger was frequently fired and the robot stayed in the vicinity of the gas source. During that period, the average distance to the gas source was approximately 1.9 m. In the trial that is shown on the right side of Fig. 7.5, an active gas source was placed at "Pos3" and a similar result was obtained. After being triggered first, the giant "robotic moth" succeeded in staying near the gas source, while the average distance to the gas source was approximately 1.8 m during that period.

For the gas source positions used in the trials shown in Fig. 7.5, the mobile robot was able to move around the source. In such cases, it finally ended up staying on one

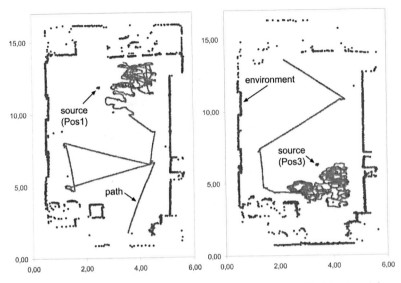

Figure 7.5: *Two experiments with an active source placed at "Pos1", started from "Start1" (left side) and with a source at "Pos3", started from "Start2" (right side). The position data concerning the environment and the path driven were computed with the scan matching algorithm of Biber et al. [BS03], using the logged laser scans and odometry data.*

side of the source, probably indicating the existance of a plume-like gas distribution oriented towards that side. If the source was placed at "Pos4", as in the trial shown on the left side of Fig. 7.6, it was not possible to drive around it. Here, the robot drove past the source at first and "found" it after exploring the room for approximately 15 minutes. Afterwards, however, it stayed in the vicinity of "Pos4" for a period of 35 minutes also residing within a small area that might indicate a plume-like gas distribution. Then, the robot departed for another 15 minutes of exploration after which the source was found again and the robot managed to stay near it during the remaining 50 minutes of the trial. A similar behaviour was also obtained in the other experiments with an active gas source shown in Fig. 7.8.

Statistical Evaluation

The experiments suggest that the modified moth strategy is able to accomplish the intended task. In order to quantify this result, the average distance to the centre of the gas source was determined for six trials with an active source and three reference

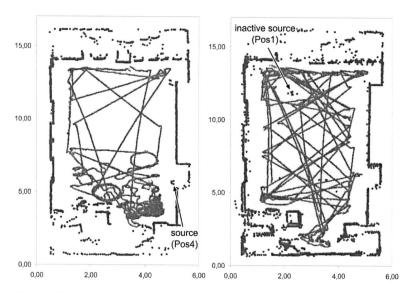

Figure 7.6: *A "robotic moth" experiment with an active source placed at "Pos4", started from "Start1" (left side) and a reference experiment with an inactive source at "Pos1", started from "Start1" (right side).*

trials. During the reference trials no gas source was used, but the wire frame was present as an obstacle at the positions "Pos1" and "Pos2". An example of such a reference experiment with the source placed at position "Pos1" is shown on the right side of Fig. 7.6.

Two different methods were applied to evaluate the experiments with an active gas source. First, the statistical quantities were calculated over the whole duration of an experiment. In order to eliminate the influence of the initial search period, only the distance values that were obtained after the fixed motion pattern was triggered for the first time were considered in a second evaluation method.

The triggering rate provides information about the expected distance to a gas source. This can be seen in Fig. 7.7, which refers to the trial that is shown on the left side of Fig. 7.6. Figure 7.7 shows the trace of the distance between the centre of the robot and the centre of the gas source (top) as well as the triggering rate (bottom), which is given by the number of restarts of the fixed motion pattern per minute. As in the trial shown in Fig. 7.7, a high triggering rate is often accompanied by a low robot-to-source distance, while prolonged periods without triggering indicate exploration phases and thus usually correspond to a high distance to the source.

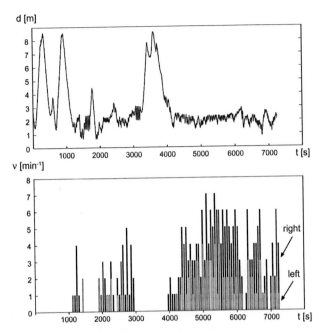

Figure 7.7: *Trace of the robot-to-source distance (top), and the triggering rate (bottom) during the trial shown on the left side of Fig. 7.6.*

The evaluation results for the experiments with and without an active source are summarised in Table 7.1. Aside from the average distance and the corresponding standard deviation, the median distance to the gas source is also given, and the values computed with both evaluation methods are itemised. In addition, the total duration that was considered for evaluation is specified.

experiment	total duration	average distance	median distance
active source, all	372.6 min	(225.6 ± 179.6) cm	266.2 cm
active source, after first trigger	297.3 min	(195.5 ± 119.8) cm	170.9 cm
inactive source	221.6 min	(317.7 ± 235.2) cm	336.8 cm

Table 7.1: *Statistics of 6 trials with an active source and 3 reference trials with an inactive source. The different methods to evaluate the trials with an active source present are explained in the text.*

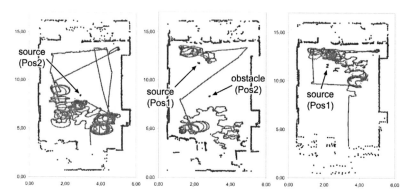

Figure 7.8: *Further "robotic moth" experiments with an active gas source, started from "Start1".*

The trials where an active source was present and the reference trials were compared by means of statistical tests. Assuming a normal distribution with identical variance, the null hypthesis of a Student t-test (H_0: both populations have identical mean) could not be rejected significantly neither if all the distance values are considered ($p_{H_0} = 0.433$) nor just those after the motion pattern was first triggered ($p_{H_0} = 0.1725$). The same holds for the distribution-free Wilcoxon test (H_0: both populations have identical median values) that is rejected with $p_{H_0} = 0.1667$ regardless of the applied evaluation method.

Due to the relatively small number of experiments as well as the limited space, a statistically validated statement is thus not possible (i.e., no conclusions can be drawn at the 95 % confidence level). The qualitative analysis of the experiments, however, shows good evidence that the general mechanisms of the moth's strategy for tracing a gas source can be applied to a considerably larger robot. Even without information about the wind direction, the robotic moth was apparently able to stay in the vicinity of a static gas source using the adapted strategy.

7.8   Conclusions

This chapter introduces a biomimetic gas source tracing strategy, which was adapted for use on a mobile robot based on the behaviour of the silkworm moth *Bombyx mori*. In the same way as the moth, the robot uses a combination of a fixed motion pattern and a triggering mechanism that (re-)starts the motion sequence if an increased concentration is sensed. The programmed motion pattern carries out a local search for the next patch of gas in terms of a zigzag movement, which is oriented against the instantaneous upwind direction in the case of the moth. Due to the detection limits of currently available anemometers, information about the local air flow direction is not available for a mobile robot in an unventilated indoor environment. Therefore, a strategy that uses only gas sensor readings had to be applied. This was achieved by omitting those parts of the moth strategy that rely on the air flow direction. Aside from the absence of a device to measure the air flow direction, the considerably larger size of the robot is the main difference compared to the biological system.

An implementation of the proposed algorithm was tested in a largely uncontrolled indoor environment and the results were evaluated by statistical means. Due to the relatively small number of experiments as well as the limited space, the obtained confidence level was not strong enough to justify a statistically significant statement concerning the superiority of the proposed method compared to random exploration. However, the results suggest that the modified strategy decreases the average robot-to-source distance compared to random exploration, because it can keep the robot in the vicinity of a gas source after single gas patches have been discovered by initial exploration.

While the adapted moth strategy made the robot stay near to the gas source, it did not effect a movement that ended directly in front of it. Considering the clearance of 85 cm, it can be said that on average a distance of approximately one metre remains in order to locate the exact position of a gas source, after the potential vicinity to a gas source has been detected. It is an advantage of the suggested algorithm that the potential vicinity to a gas source is indicated by a continued period of high triggering frequency. Thus, it provides a very rough gas source declaration method. In order to achieve a more accurate localisation of the gas source, the adapted moth strategy would have to be extended by a supplementary strategy to trace the gas source after it is approximately located with the adapted *Bombyx mori* algorithm, and a more accurate mechanism to detect close proximity to the gas source. This problem is addressed in Chapter 8.

Chapter 8

Learning to Detect Proximity to a Gas Source

"To be, or not to be: that is the question." (Shakespeare, "Hamlet")

This chapter is concerned with the, conceptually speaking, final step of gas source localisation: the task of determining the certainty that a gas source is in the immediate vicinity in order to terminate the search (gas source declaration). It presents first results of an ongoing study that investigates the classification performance, which can be achieved based on gas sensor readings only.

8.1 Gas Source Declaration

The ability to classify an object depending on whether it is a source of gas or not can be useful for mobile robots for several reasons. First, it is an essential part of gas source localisation. Second, the classification capability of gas source declaration itself is of potential use for rescue and security missions even if the full gas source localisation problem cannot be accomplished using a sense of smell only (because of too low concentration at locations distant from the source, for example). An object that is to be classified could be located using other sensor modalities, and attributed based on gas sensor measurements. For example, suspicious items could be identified as containing explosive materials or a rescue robot could determine whether a victim is alive by assessing whether that person is a source of carbon dioxide. Note that CO_2 emmision belongs also to the characteristics defined in the RoboCup Rescue scenario [KTN+99] by which the simulated victims display signs of life. While in rescue scenarios other sensor modalities will also be used to check for vital signs [CM02], a mobile robot that is equipped with gas sensors would be able to monitor the possibly contaminated air at an emergency site. Thus, the robot could prevent rescue workers from being harmed or killed due to explosions, asphyxiation or toxication [MCH+00]. Furthermore, such a rescue robot could assemble a map of the spatial gas distribution (see Chapter 5), providing incident planning staff with information to support rational decision making.

8.1.1   Gas Source Declaration Using Gas Sensors Only

Throughout the animal kingdom, many examples can be found where olfactory information plays an important role for performing different localisation tasks. These tasks include finding a mate guided by sexual pheromones, like the male silkworm moth *Bombyx mori* for instance, or locating a host as is painfully known from bloodsucking mosquitoes, which are attracted by a specific mixture of human scents, often including carbon dioxide [PCLL+01] and lactic acid [DSCG02]. Host-oriented behaviour of animals however, often relies on a combination of cues determined from a variety of sensor modalities. Mosquitoes and other flying haematophagous insects are also attracted by particular colours [GT99] (especially blue, which represents a strong stimulus meaning "non-vegetation"), while the *Bombyx* males use the local wind direction as an approximation to the direction of a pheromone source (see Chapter 7 and [Kan96; Kan98b]). The same is probably true for the mechanisms applied to identify a host.

This chapter is concerned with the classification performance that can be achieved using gas sensors only. In an environment with a strong air flow, it should be beneficial to use additional information about the local wind speed. Here, however, an indoor environment without strong artificial ventilation is considered where it is not feasible to reliably measure the typically weak air currents due to the detection limits of current anemometers (see Section 2.8).

The approach that is suggested here does not depend on sufficiently high wind speeds. It tries to classify the inspected object by recognising a pattern within a series of gas sensor readings that represent temporally as well as spatially sampled concentration data. It is understood that the small-scale structure of a concentration field contains information about the distance to the source [Ate96; BS02]. In an environment with a unidirectional airflow, proximity to a gas source could be further determined by detecting the decreasing width of the gas plume from the fluctuating sensor signal. Thus, it should be possible to find a pattern that enables discrimination of situations where the robot is in the immediate vicinity of a source from those where the series of sensor readings is recorded at a distance from a gas source. Such a pattern is determined in this work by applying machine learning techniques to a set of experiments carried out in an uncontrolled indoor environment. To the authors' knowledge there is no physically justified model available yet to establish the required pattern in the case of a natural environment by analytical means.

It is important to recall the properties of a gas distribution in a real world indoor environment without a strong unidirectional airflow discussed in Section 2.2. Due to the low diffusion velocity, the dispersal of an analyte gas is dominated by turbulence and the prevailing air flow rather than diffusion. The gas distribution therefore reveals many discontinous patches and the absolute maximum of the instantaneous distribution is usually *not* located near the gas source if this source has been active for some time. It is therefore not sufficient to search for maxima of the instantaneous concentration distribution in order to solve the gas source declaration task.

8.2   Related Work

Several suggestions for gas source declaration are discussed in Section 2.11. If detailed information about the air flow and the intensity of a gas source is available, the distance to the source could be determined from time-averaged concentration measurements [RBHSW03]. Previous knowledge about the features of a gas source that appear to other sensor modalities might also assist the declaration step. However, this information will often not be available in a typical application scenario. It is thus desired to utilise more general characteristics to determine proximity to a gas source. In an environment with a sufficiently strong unidirectional airflow, a gas source could be identified by determining a concentration drop on the upwind side of the source [RTDMS95; HMG02] or by detecting a reducing plume width while approaching the source [RBHSW03]. However, because detailed experimental evaluations are not available, it remains an open question as to what amount of temporal averaging is necessary to extract these characteristics from the turbulent concentration field. An interesting alternative would be to utilise positional clues in the fine structure of a turbulent distribution [Ate96; RBHSW03]. With the exception of an implementation by Russell et al. [RTDMS95] where the robot determined whether the source was found by detecting a concentration drop on the upwind side of the source, none of the mentioned solutions to the problem of gas source declaration were verified experimentally so far. While a few experiments have been published where a gas source localisation strategy was applied that includes the declaration step (as in the model-based approaches discussed in Section 2.10), no evaluation of the corresponding declaration performance by statistical means is available either.

A further solution to the problem of gas source declaration that falls in the last defined category is provided by the reactive localisation strategy based on exploration and concentration peak avoidance discussed in Section 6.5.2. Here, a gas source was located by exploiting the fact that local concentration maxima occur more frequently near the gas source compared to distant regions. The concentration mapping technique introduced in Chapter 5 provides another possibility for gas source declaration. As discussed in Section 5.9, the position of the maximum in the representation of the average relative concentration of a detected gas can often be used to estimate the approximate location of the source. However, the latter two approaches suffer from similar drawbacks. Aside from an increased time consumption (though this can be reduced by using multiple robots) it is not guaranteed that a good estimate of the source location can be obtained with these techniques and there is yet no method available to determine the certainty of this estimate.

This chapter introduces a direct declaration strategy that tries to determine whether a gas source is located in the immediate vicinity of the robot from a series of concentration measurements, recorded while the robot performed a rotation manoeuvre in front of a possible gas source (see Section 8.4 for details). A similar approach was used by Duckett et al. [DAS01] to learn the direction to a gas source from a series of sensor readings (see Section 2.8).

The work presented in this chapter is concerned with the task of gas source declaration while the gas source tracing problem is not addressed here. It is rather assumed that the source appears as an obstacle to the robot, which has to be analysed after it has been detected using other sensor modalities (e.g., vision or range-finder sensors).

The rest of this chapter is structured as follows: next, the experimental setup is described in Section 8.3 and the applied declaration strategy is introduced in Section 8.4. Then, the pre-processsing of data is detailed (Section 8.5) and corresponding results are discussed (Section 8.6), followed by conclusions and suggestions for future work (Section 8.7).

8.3   Experimental Setup

8.3.1   Robot and Gas Sensing System

The gas source declaration strategy that is introduced in Section 8.4 was implemented on the gas-sensitive robot "Arthur" shown in Fig. 3.7. For the experiments presented in this chapter only odometry data were utilised in addition to the concentration measurements. The data from the SICK laser range scanner were used to determine the position of the robot for evaluation purposes.

The gas sensing system is based on the commercially available device VOCmeter-Vario. It is described in detail in Section 3.4.2. For the gas source declaration experiments seven metal oxide sensors (see Section 2.1) were utilised, which were placed on the robot as shown in Fig. 8.1. Five TGS 2620 sensors were symmetrically mounted at a height of 9 cm above the floor on the front bumper of the robot. The

No.	Position	Sensor Type	Coordinates
0/4	•° ∘∘∘ °•		(40 cm, ∓32 cm, 9 cm)
1/3	∘° •∘• °∘	TGS 2620	(40 cm, ∓16 cm, 9 cm)
2	∘° ∘•∘ °∘		(40 cm, 0 cm, 9 cm)
5/6	∘• ∘∘∘ •∘	TGS 2600	(40 cm, ∓22 cm, 16 cm)

Table 8.1: *Sensor positions utilised in the gas source declaration experiments with respect to the centre of the robot. The second column shows an iconic visualisation of the sensor position corresponding to a front view of the robot (see Fig. 8.1).*

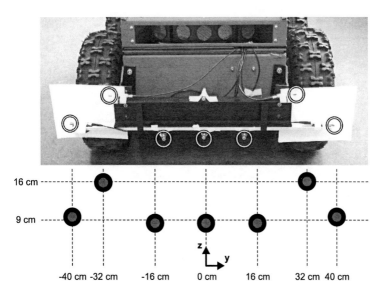

Figure 8.1: *Setup of the sensor array used in the gas source declaration experiments.*

distance of these sensors to the middle of the bumper was 0 cm, ± 16 cm, and ± 40 cm. Two additional sensors of type TGS 2600 were mounted at a height of 16 cm with a distance of ± 32 cm to the centre. Table 8.1 itemises the sensor positions with respect to the centre of the robot and gives a visualisation of their relative locations by means of an iconic front view of the robot.

The distance between the outer sensors (especially sensor 0 and sensor 4) and the front wheels is very small. In order to avoid a corruption of the results due to an additional airflow created by the wheels, a shield made of cardboard was placed inbetween the wheels and the sensors (see Figs. 8.1 and 3.7).

8.3.2 Environment and Gas Source

All experiments were carried out in a 15.4 m × 5.1 m room at the University of Tübingen. A floor plan is shown in Fig. 8.2, including doors, windows, cupboards and desks. In addition, the tested gas source positions are indicated by circles. A total of $N = 1056$ declaration trials were performed using three different source locations and four different orientations with respect to the source as indicated in Fig. 8.2. For each source position, 176 experiments were carried out at a distance d directly in front of the gas source ($d = d_0$) alternating with 176 trials at a randomly

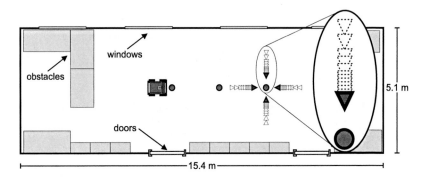

Figure 8.2: *Floor plan of the laboratory room in which the gas source declaration experiments were performed. The tested locations of the gas source are indicated by circles. Beneath the source on the left side, the robot is sketched in a position that was considered as being directly in front of the source. For the source location on the right, all the tested robot positions are shown using triangles that indicate the centre of the robot and its initial heading. Triangles with a dotted border indicate positions that were considered as being not in the immediate vicinity of the source.*

chosen larger distance of $d = d_0 + \Delta d$ with $\Delta d = 5$ cm, 10 cm, 15 cm, 20 cm, 25 cm, 30 cm, 40 cm, 50 cm, 60 cm, 80 cm and 100 cm, respectively. After each trial, the robot was stopped for 60 s in order to avoid disturbance from the preceding measurements due to the long decay time of the sensors. All the robot positions tested are shown in Fig. 8.2 for the right source position, using triangles that indicate the centre of the robot and its initial heading. Light triangles with a dotted border indicate positions that were considered as being not in the immediate vicinity of the source. With regard to real world applications, the environment was not modified for this investigation. The unventilated room was also used as an office during the experiments, with up to two persons working, moving and sometimes leaving or entering the room. Although the windows were kept closed and the persons were told to be careful, this indoor environment can be considered uncontrolled to some extent.

The gas source was chosen to be a bowl with a diameter of 140 mm and a height of 20 mm filled with Single Malt Whiskey (40% alcohol), which was used because it is non-toxic, less volatile than pure ethanol and easily detectable by metal oxide sensors. In order to be recognisable by the laser range scanner, a frame made of wire with a cardboard marking mounted on top was placed above the container (see Figs. 3.7 and 7.2). The container was filled and covered with a plastic sheet approximately 1 hour before a series of experiments was started. Then, the room was ventilated by opening the windows for 30 minutes, and temperature and airflow

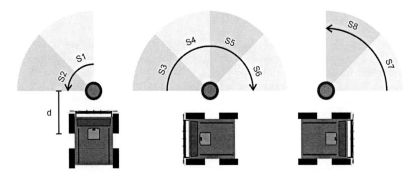

Figure 8.3: *Rotation manoeuvre performed to collect sensor data for gas source declaration. The manoeuvre is displayed in three steps corresponding to the three successive rotations executed. Each picture shows the position of the robot before the corresponding rotation is carried out, the gas source, and the sectors for which the mean and standard deviation is calculated as a feature for input to a pattern recognition system.*

disturbances were compensated for by waiting another 20 minutes after the windows had been closed. Finally, the sheet was removed and the experiment was started 10 minutes later.

8.4 Gas Source Declaration Strategy

Due to the properties of gas distribution in real world environments discussed in Section 2.2, single concentration measurements do not contain enough information to declare the vicinity to a gas source. It was instead considered most promising to apply a strategy that provides temporally as well as spatially sampled concentration data. Therefore, the gas sensor readings were acquired while the robot performs a rotation manoeuvre containing three successive rotations: 90° to the left, then 180° to the right (without stopping, in order to minimise self-induced disturbance of the gas distribution) and finally 90° to the left again (see Figure 8.3). Initially, the robot was oriented towards the suspected object as indicated in Fig. 8.3. This manoeuvre is easy to implement, requires little space and does not involve periods of backward motion where the ATRV-Jr robot offers only a limited obstacle avoidance capability. The rotation was performed with an angular speed of approximately 4°/s corresponding to a total time of approximately 90 s to complete the manoeuvre. Simultaneously, sensor readings were acquired at the maximum rate of approximately 4 Hz, resulting in a total of Q readings per experiment with $Q \in [349, 362]$.

8.5   Data Pre-Processing

To evaluate the performance of the two machine learning methods tested (artificial neural network and support vector machine), the recorded data were first preprocessed by means of feature extraction (Section 8.5.1) and normalisation (Section 8.5.2). Next, a desired output value was added to each training example, indicating whether the corresponding experiment was performed directly in front of a gas source $(+1)$ or not (-1). The robot was considered as being in the "immediate vicinity of a source" only in the case of minimal distance between the robot and the gas source, corresponding to a laser scanner reading of $d = d_0 = 50$ cm (see Fig. 8.3). Here, the trajectory of the sensors just avoids hitting the object under inspection at the point of closest approximation. By contrast, all the positions with a larger distance $d > d_0$ were considered as being "not in the immediate vicinity of a source". Thus, the distance of these trials was $d \geq d_0 + \Delta d_{min}^{ns}$ with the minimal distance of negative examples from the point of closest approximation Δd_{min}^{ns}.

8.5.1   Features

The features used for classification were derived by calculating the first two statistical moments (mean and standard deviation of the sensor measurements) for each of the 8 consecutive 45° sectors covered by the rotation manoeuvre. These sectors are denominated by S1 – S8 in Figures 8.3 and 8.6. Depending on the number M of gas sensors utilised, a maximum of $M \times 16$ features was extracted. Either the full $M \times 16$-dimensional input vector was utilised for training and testing, or only the $M \times 8$ mean or standard deviation values.

In order to get an idea of the expected feature patterns, the theoretical response of a perfect sensor to a time-averaged gas distribution can be considered. The effects of turbulence can be described on average as diffusion in specific situations [INM98]. Assuming isotropic and homogenous turbulence and a one-directional wind field with a possibly non-zero wind speed that is constant on average, the time-averaged gas distribution of a point source on the floor can be described as [INM98; Hin75]:

$$C_0(x, y) = \frac{q}{2\pi K} \frac{1}{r} exp[-\frac{V}{2K}(r - \Delta x)], \tag{8.1}$$

$$r = \sqrt{(x_s - x)^2 + (y_s - y)^2}, \tag{8.2}$$

$$\Delta x = (x_S - x)cos(\theta) + (y_S - y)sin(\theta). \tag{8.3}$$

The concentration C_0 at a point (x, y) on the floor is determined by the turbulent diffusion coefficient K, the location of the gas source (x_S, y_S) and its release rate q as well as the wind speed V and the upwind direction θ.

In order to calculate the concentration divided by the amount of analyte gas released per time unit $(C_{rel} = C_0/q)$ with Eqs. 8.1 – 8.3, three model parameters have to be estimated. Apart from the wind direction and speed, this applies also to

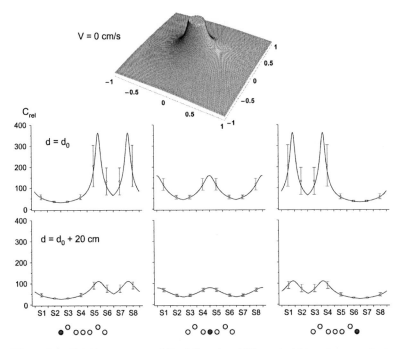

Figure 8.4: *Relative response of the leftmost, middle and rightmost sensor during the rotation manoeuvre, assuming perfect sensors and an isotropic gas distribution. Note that the graph at the top of the figure, which indicates the radially symmetric distribution, was cropped for clarity of the plot.*

the eddy diffusivity K. For the considered environment, the wind speed V can be estimated with a value lower than 5 cm/s, while different wind directions have to be considered in the case of a non-zero value. The eddy diffusivity is approximated as $K = 0.005$ m^2/s, which corresponds to a mean square displacement due to turbulence that is approximately 20 times higher compared to the molecular diffusion of gases at room temperature. (The diffusion constant for ethanol vapour in air at room temperature is $D_{eth} = 0.119$ cm^2/s [NIM99]).

Fig. 8.4 shows the response of perfect sensors according to Eq. 8.1 in the case of an isotropic distribution ($V = 0$ cm/s). The graphs in the lower part display the relative concentration in three columns, according to the position of the leftmost, middle and rightmost sensor during the rotation manoeuvre, and two rows, according to two distances d_0 and $d_0 + 20$ cm from the source. It is understood that the point

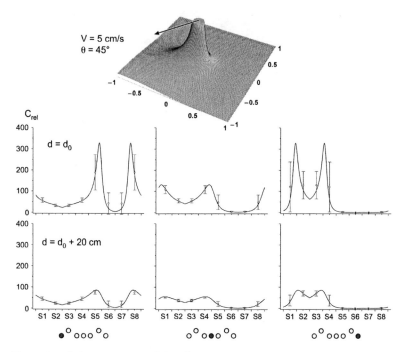

Figure 8.5: *Relative response of the leftmost, middle and rightmost sensor during the rotation manoeuvre, assuming perfect sensors and a non-isotropic gas distribution that is caused by an average wind speed of 5 cm/s from $\theta = 45°$. Note that the graph at the top of the figure, which indicates the radially symmetric distribution, was cropped for clarity of the plot.*

source is located at $(x_S, y_S) = (0 \text{ cm}, 0 \text{ cm})$ and the robot at $(0 \text{ cm}, -d_0)$ and $(0 \text{ cm}, -d_0 - 20 \text{ cm})$ with $d_0 = 60$ cm, resulting in a minimal distance between the sensors and the gas source of 8.8 cm. In addition to the concentration, the mean and standard deviation values computed for the eight sectors S1 – S8 are also displayed by eight dots with error bars plotted on top of the sensor response.

According to the applied model, the relative course of the mean values exhibits little variation with varying distance from the source, but the absolute height of the sensor response decreases quickly with an increasing distance. A similar behaviour is predicted by the applied model in the case of moderate wind speeds. This can be seen in Fig. 8.5, which shows the expected sensor response in the case of $V = 5$ cm/s and $\theta = 45°$ in the same way as Fig. 8.4.

The quickly decreasing absolute height of the response suggests that it should be feasible to solve the gas source declaration problem based on a threshold that is applied to the mean response. Therefore, the classification performance that is achievable by selecting an optimal threshold value regarding the mean sensor signal during the rotation manoeuvre was also tested as a possible method for gas source declaration (a so-called "threshold classifier"). However, apart from the problem of sensor drift due to changing environmental conditions or ageing of the sensors (see Section 2.1), the suitability of the threshold classifier is limited by the fact that a weak sensor response occurs in the case of non-zero wind speeds also if the robot is located upwind from the source. Moreover, the signal obtained from a real distribution is superimposed by strong local concentration variations due to turbulence (typically with high peak-to-mean ratios of 10 or more [RW02]) and depends also on the time since the source was uncovered because a non-stationary situation is considered.

Examples of feature vectors obtained in the experiments are depicted in Fig. 8.6. Here, the vertically normalised mean values of the leftmost, middle and rightmost sensor are plotted in order to indicate the relative strength of the sensor responses. Four examples are depicted for each of the four distances d_0, $d_0 + 20$ cm, $d_0 + 40$ cm, and $d_0 + 80$ cm to the gas source.

8.5.2 Normalisation

The set of feature vectors $\vec{F}_i$ (corresponding to the desired classification t_i of the i-th experiment) creates a matrix F_{ij} ($j \in [1, M \times 8]$ or $j \in [1, M \times 16]$ and $i \in [1, N]$ with the number of experiments N and the number of sensors M). Before training and testing, this matrix is normalised *vertically*, meaning that each column is mapped linearly to the range of [0,1] as

$$f_{ij}^v = \frac{F_{ij} - min\{F_{\bullet j}\}}{max\{F_{\bullet j}\} - min\{F_{\bullet j}\}}. \tag{8.4}$$

Note that this kind of normalisation cannot be applied in the same way for classification of a single trial because it is necessary to know all N experiments in order to establish the normalisation range. It might be also problematic to apply the normalisation factors obtained from the training data in a test experiment in the case of varying environmental conditions that cause a shift of the sensor values, such as a different temperature or humidity. Finally, the vertical normalisation factors contain knowledge about the intensity of the gas source used in the training phase, and could thus be misleading in the case of a different source.

For online evaluation of single experiments *horizontal* normalisation could be used:

$$f_{ij}^h = \frac{F_{ij} - min\{F_{i\bullet}\}}{max\{F_{i\bullet}\} - min\{F_{i\bullet}\}}. \tag{8.5}$$

While in the case of vertical normalisation, the available information about the strength of the sensor response (relative to the range experienced in all the exper-

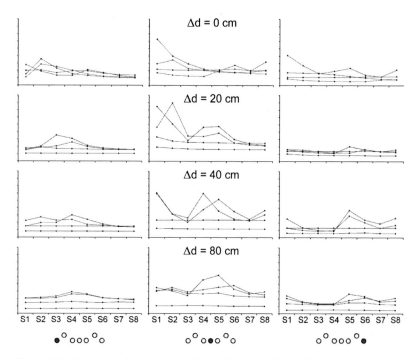

Figure 8.6: *Examples of mean values obtained at the indicated distance from the gas source. For each distance, four vertically normalised feature vectors are shown.*

iments) is included in the feature vector, a horizontally normalised feature vector represents the relative intensity of the sensor response with respect to the values that occur during the rotation manoeuvre. Therefore, examples have to be classified in the latter case based on the relative course of the concentration measurements only. For real world applications, however, the concentration measurements collected before the rotation manoeuvre started can also be used to acquire an approximation of the range that is used for vertical normalisation. The results obtained with horizontal and vertical normalisation provide therefore a lower and upper boundary of the classification performance that can be achieved in real world applications where the robot collects gas sensor readings on its way to inspected objects.

Instead of using the scaling given in Eqns. 8.4 and 8.5, the linear transformation that normalises the components of the feature vectors to have zero mean and unit standard deviation [Bis95] was also tested (either vertically and horizontally). Because no improvement of the classification performance was obtained using this

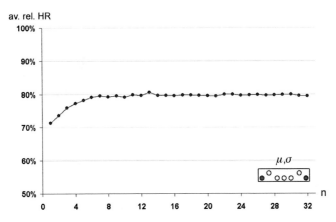

Figure 8.7: *Total hit rate of a multilayer feedforward network for $\Delta d_{min}^{ns} = 60$ cm ($\overline{\Delta d^{ns}} = 80$ cm) depending on the number n of hidden neurons. The input layer contains 32 neurons corresponding to the mean and standard deviation of the readings of the left- and rightmost sensor.*

transformation, only feature vectors that are normalised according to Eqns. 8.4 and 8.5 are considered for the remainder of this chapter.

8.6 Results

At each of the gas source positions indicated in Fig. 8.2, four experiments were carried out at four different directions (north, east, south, west) and eleven different distances $\Delta d > 0$, alternating with four experiments in the direct vicinity of the gas source ($\Delta d = 0$). Thus, a total of $N = 3 \times 4 \times 4 \times 11 \times 2 = 1056$ declaration trials were performed including 528 trials (50%) in the immediate vicinity of the source ($d = d_0$) and 528 experiments (50%) at a larger distance of $d = d_0 + \Delta d$ with $\Delta d \geq 5$ cm. Using the obtained data set, the two pattern recognition algorithms were evaluated by means of 5-fold cross-validation. In order to increase the accuracy of the evaluation, the hit rate (the percentage of correctly classified examples) was calculated by averaging over fifteen 5-fold cross-validation runs.

8.6.1 Artificial Neural Network

One way to solve non-linear classification problems is provided by artificial neural networks. Here, a multilayer feedforward (MLFF) network with a sigmoidal activation function was used, containing an input layer with $N_{in} = M \times 8$ or $M \times 16$

neurons, an output layer of one unit and a hidden layer with a variable number of $n \in [1, N_{in}]$ neurons. The achieved classification rate, however, was found to reach an approximately stable level below $n \approx 8$ hidden neurons. This can be seen in Fig. 8.7, which shows the hit rate depending on the number of hidden neurons using the mean and standard deviation of the two outermost sensors as a 32-dimensional feature vector. Training was performed using conjugate gradient descent [RM86] with 100 training cycles and a weighting of $\alpha = 0.1$ for the momentum term, which was found to yield good results in initial tests. The weights were initialised with a randomly chosen set of values at the beginning of each run.

8.6.2   Support Vector Machine

Support Vector Machines (SVM) [CV95] have become established during the recent years as a major state-of-the-art classification method. The main idea is to construct a so-called optimal separating hyperplane between two classes A and B, which maximises the margin between the convex hulls of A and B. It is understood that a larger margin leads to a better generalization performance of the SVM [Vap95]. Learning corresponds to a dual optimisation problem, for which a unique and global optimal solution can be obtained by quadratic programming. If A and B are not linearly separable, the so called *kernel trick* is applied. All input patterns are mapped via a nonlinear function Φ into a high dimensional feature space where the problem is linearly separable again. It is, however, usually not necessary to perform this transformation directly. Instead, the *kernel function* k can be used, which can be thought of as a similarity measure between two patterns x and y, representing a dot product in feature space. As a popular kernel function the radial basis function is used here as

$$k_\gamma(x, y) = \exp(-\frac{\|x - y\|^2}{\gamma^2}). \tag{8.6}$$

In order to find suitable learning parameters, a grid search was carried out in the two-dimensional search space spanned by the kernel parameter γ and the parameter C that determines the extent to which outliers are penalised. At this, 2205 points were sampled for each set of feature vectors at $\gamma = 2^{-3}, 2^{-2.75}, ..., 2^8$ and $C = 2^{-6}, 2^{-5.75}, ..., 2^6$, which covers the parameter range where all the optimal combinations were found in initial tests.

Corresponding results are itemised in Table 8.2, which shows a comparison of the classification performance achieved with the artificial neural network and the support vector machine. The first two columns specify the used feature vector, including the considered sensors and statistical moments (mean "μ" and/or standard deviation "σ"). Then, the best result in terms of maximum hit rate (HR) obtained with a MLFF neural network (third and fourth column) and the support vector machine (fifth and sixth column) is given. In addition to the maximum hit rate, the corresponding average cross-validation rate of false positives (FP) and false negatives (FN) is also itemised. Finally, the corresponding parameters for which the best classification performance was achieved (the number of hidden neurons n^* in the case of the MLFF network and the penalty and kernel parameter (C^*, γ^*) in the case of support vector machines) are also given. The results in Table 8.2 were obtained with a subset of the data containing $N = 288$ examples, which was created from the

vertical normalisation					
$\Delta d_{min}^{ns} = 60$ cm		MLFF neural network		SVM	
Sensors	Features	n_v^*	HR (FP, FN)[%]	C_v^*, γ_v^*	HR (FP, FN)[%]
○ ○ ●○ ○ ○	μ (8)	6	75.7 (22.5, 26.1)	$2^{1.25}, 2^{3.25}$	77.1 (24.5, 21.3)
	σ (8)	8	75.5 (20.3, 28.7)	$2^{-0.25}, 2^{4.75}$	77.1 (25.4, 20.3)
	μ, σ (16)	9	75.6 (22.5, 26.3)	$2^{8.25}, 2^{4.5}$	78.1 (13.6, 30.2)
● ○ ○○○ ○ ●	μ (16)	16	76.8 (19.9, 26.4)	$2^{4.25}, 2^4$	80.0 (15.1, 24.9)
	σ (16)	13	76.3 (20.4, 27.0)	$2^{3.25}, 2^1$	78.3 (26.4, 17.1)
	μ, σ (32)	13	80.6 (19.5, 19.4)	$2^{3.75}, 2^{-0.25}$	83.0 (19.0, 15.0)
● ○ ○●○ ○ ●	μ (24)	16	82.7 (16.3, 18.4)	$2^{5.5}, 2^{2.5}$	**87.5 (10.3, 14.8)**
	σ (24)	23	80.8 (15.6, 22.7)	$2^5, 2^{0.5}$	83.4 (18.7, 14.5)
	μ, σ (48)	24	**84.1 (14.9, 16.9)**	$2^{6.75}, 2^{0.5}$	86.4 (10.1, 17.2)
horizontal normalisation					
○ ○ ●○ ○ ○	μ (8)	1	61.0 (24.6, 53.4)	$2^{2.25}, 2^{-1.5}$	64.1 (32.5, 39.3)
	σ (8)	7	60.4 (38.7, 40.5)	$2^{1.75}, 2^{3.25}$	63.4 (36.3, 36.9)
	μ, σ (16)	14	62.7 (36.3, 38.2)	$2^{0.75}, 2^3$	66.0 (40.5, 27.5)
● ○ ○○○ ○ ●	μ (16)	12	68.1 (31.6, 32.8)	$2^{3.75}, 2^1$	70.3 (30.2, 29.3)
	σ (16)	16	60.3 (39.7, 39.7)	$2^{2.75}, 2^{-0.75}$	65.5 (39.0, 30.1)
	μ, σ (32)	25	71.5 (27.0, 30.0)	$2^{4.5}, 2^{-2.75}$	73.0 (26.7, 27.4)
● ○ ○●○ ○ ●	μ (24)	24	69.0 (31.1, 30.9)	$2^3, 2^{0.5}$	74.9 (28.8, 21.3)
	σ (24)	20	63.7 (34.5, 38.0)	$2^6, 2^{-2.75}$	70.2 (34.6, 25.0)
	μ, σ (48)	36	**73.7 (24.9, 27.7)**	$2^{2.25}, 2^{-2}$	**75.7 (24.6, 24.0)**

Table 8.2: *Comparison of the classification performance obtained with a MLFF network and the SVM for $\Delta d_{min}^{ns} = 60$ cm $(\overline{\Delta d^{ns}} = 80$ cm). Apart from the average cross-validation performance in terms of the hit rate (HR), the rate of false positives (FP) and the rate of false negatives (FN), the optimal learning parameters are also given. For the MLFF network this is the number of hidden neurons n^* and for the SVM the penalty parameter C^* and the kernel parameter γ^*. The index "v" or "h" indicates whether the feature vectors were vertically or horizontally normalised.*

full data set by disregarding those trials with 0 cm $< \Delta d < \Delta d_{min}^{ns} = 60$ cm. Thus, the distance of negative examples was $\Delta d = 60$ cm, 80 cm or 100 cm, resulting in an average distance of $\overline{\Delta d^{ns}} = 80$ cm. Because the classification performance obtained with the MLFF neural network was lower for all configurations tested, only support vector machines were considered for the evaluation of the dependency between the hit rate and the mean distance of negative examples from the source.

8.6.3   Dependency on the Distance From the Source

In order to assign the training examples to different categories according to whether they are recorded in the immediate vicinity of the gas source or not, the examples were separated by the distance to the source at which the rotation manouevre was performed. Positive examples were assumed when the data were collected at the minimal distance d_0 where the trajectory of the gas sensors just avoids hitting the source at the point of closest approximation. Negative examples were assumed when the data were collected at a larger distance of $d \geq d_0 + \Delta d_{min}^{ns}$. The correlation between the classification performance and the mean distance of negative examples to the source was investigated by disregarding those trials with $0 < \Delta d < \Delta d_{min}^{ns}$ for evaluation. In order to preserve an even proportion between positive and negative

N	Δd_{min}^{ns}	$\overline{\Delta d^{ns}}$	vert. normalisation HR (FP, FN) [%]	hor. normalisation HR (FP, FN) [%]	threshold HR (FP, FN) [%]
96	100 cm	100.0 cm	97.1 (5.4, 0.4)	82.6 (16.1, 18.7)	79.8 (27.6, 12.8)
192	80 cm	90.0 cm	93.7 (5.6, 7.0)	82.5 (17.6, 17.5)	80.9 (21.2, 17.0)
288	60 cm	80.0 cm	88.9 (9.6, 12.7)	79.2 (22.7, 19.0)	75.5 (41.8, 7.0)
384	50 cm	72.5 cm	85.7 (12.4, 16.3)	76.4 (23.0, 24.2)	72.0 (41.8, 14.3)
480	40 cm	66.0 cm	85.7 (13.7, 15.0)	78.7 (22.0, 20.5)	68.0 (43.4, 20.4)
576	30 cm	60.0 cm	81.7 (14.2, 22.5)	73.8 (24.1, 28.3)	69.2 (46.0, 15.4)
672	25 cm	55.0 cm	77.5 (23.1, 21.8)	71.6 (26.9, 29.9)	64.5 (48.0, 23.1)
768	20 cm	50.6 cm	73.7 (23.7, 29.0)	67.8 (32.3, 32.1)	61.4 (48.8, 28.4)
864	15 cm	46.7 cm	71.4 (26.6, 30.6)	65.7 (34.1, 34.6)	59.7 (51.3, 29.2)
960	10 cm	43.0 cm	68.2 (29.6, 34.0)	63.1 (36.7, 37.1)	57.6 (52.1, 32.7)
1056	5 cm	39.5 cm	65.9 (38.6, 29.7)	61.1 (38.0, 39.9)	56.4 (60.6, 26.6)

Table 8.3: *Average cross-validation performance (hit rate HR, rate of false positives FP and rate of false negatives FN) for a varying average distance $\overline{\Delta d^{ns}}$ of negative examples to the gas source. The last three columns itemise the best classification rate obtained with the support vector machine (using vertical and horizontal normalisation) and by selecting an optimal threshold on the raw data ("threshold classifier").*

examples, the loss of negative examples was then compensated by omitting the same amount of randomly chosen positive examples.

Table 8.3 shows the maximum hit rate that could be obtained with the support vector machine (using vertical and horizontal normalisation) and by selecting an optimal threshold on the raw data ("threshold classifier"). In addition to hit rate (HR) the cross-validation rate of false positives (FP) and the rate of false negatives (FN) are also given in the table. The first three columns itemise the total number N of positive and negative examples used for training and testing as well as the minimal distance between positive and negative examples Δd_{min}^{ns} and the average of this distance for all the negative examples considered $\overline{\Delta d^{ns}}$. The optimal threshold was acquired with respect to the mean sensor signal during the rotation manoeuvre, meaning that the threshold classifier refers to a time-averaged concentration value. It should be noted that no cross-validation was carried out in order to determine the performance of the threshold classifier.

Because a gas source with an approximately constant intensity was utilised in all the trials, high classification rates would be expected for the threshold classifier in the case of an isotropic distribution. However, the observed intensity exhibits pronounced variations depending on the orientation of the robot with respect to the source, indicating a weak constant air flow as assumed in Fig. 8.5. The concentration drop on the upwind side leads to considerably lower intensities of many positive training examples, resulting in a reduced classification threshold. As a consequence, negative examples are often misclassified, meaning that the threshold classifier is prone to a high rate of false positives.

The general trend of the classification performance against the average distance of negative examples from the gas source can be seen in Fig. 8.8. This plot shows a comparison of the cross-validation hit rate obtained with the support vector machine using vertical normalisation and the performance of the threshold classifier. Apart from the maximum hit rate that was achieved with the support vector machine, considering all the feature vectors (also itemised in Table 8.3), individual results for a few selected feature vectors are also given. In the same way, Fig. 8.9 shows the results obtained with horizontal normalisation.

As expected, the classification performance decreases with decreasing distance to the source. The observed course, however, is not linear and three different regions can be distinguished. An approximately linear descent was found when the average distance of negative examples was above 72.5 cm (corresponding to $\Delta d^{ns} \geq 50$ cm) or below 60 cm ($\Delta d^{ns} \leq 30$ cm) with a roughly constant plateau in between. A similar profile was observed for all of the classifiers tested, probably indicating a transition between domains where proximity to the gas source can be detected using different properties of the concentration field.

The classification performance that was achieved with the support vector machine using vertical normalisation was generally higher compared to the performance of the threshold classifier. This result corresponds to the fact that information about

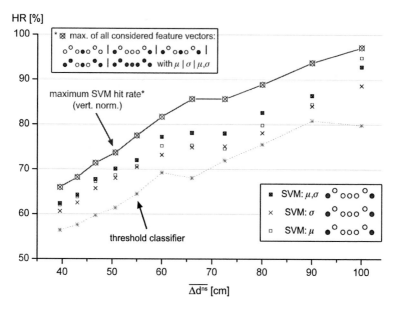

Figure 8.8: *Comparison of the classification performance obtained with the threshold classifier and the support vector machine using vertical normalisation. The achieved hit rate is plotted against the average distance of negative examples from the gas source. Different feature vectors were considered as indicated in the figure.*

the relative course of the sensor signal can be exploited by the SVM in addition to information about the absolute intensity, which is only used by the threshold classifier. In contrast to vertical normalisation, horizontal normalisation does not preserve the absolute intensity. Although the achieved classification performance was generally lower using horizontally normalised feature vectors, the maximum hit rate achieved with the SVM was nevertheless higher than the performance of the threshold classifier (see Fig. 8.9).

While using only the mean values of each sector or only the standard deviation yielded similar hit rates in the case of vertical normalisation, the standard deviation was found to be of little value with horizontal normalisation. Consequently, the maximum hit rate with vertical normalisation was most often found using both the mean values and the standard deviations, while the additional benefit of using also the standard deviations was rather small in the case of horizontal normalisation. This tendency can be seen with the unconnected {blue} symbols in Figs. 8.8 and 8.9 that indicate the results, obtained using only the two outermost sensors.

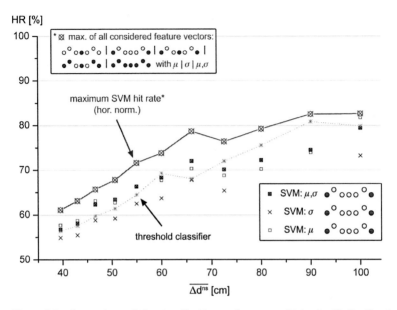

Figure 8.9: *Comparison of the classification performance obtained with the threshold classifier and the support vector machine using horizontal normalisation. The achieved hit rate is plotted against the average distance of negative examples from the gas source. Different feature vectors were considered as indicated in the figure.*

Five different sensor combinations were considered for evaluation, including one, two, three, five or seven sensors. The average relative hit rate for each of these sensor combinations is shown in Fig. 8.10 for vertical normalisation (where the symbols are connected by a solid line) and horizontal normalisation (dotted line). Here, the mean and standard deviation of each sector were used as the feature vectors. The relative hit rate for a particular sensor combination was calculated by dividing the obtained hit rate with the maximum hit rate that was achieved at a given distance Δd_{min}^{ns} considering all the five different sensor combinations. Then, the average relative hit rate was computed by averaging over all the considered distances Δd_{min}^{ns}. While the general trend observed was that using more sensors yielded better results, the improvement obtained using more than three sensors was small. Especially adding the sensors in the upper row ($z = 16$ cm) was not found to be superior compared to using only the outermost and middle sensor in the lower row. Due to the very similar response to alcoholic substances of the TGS 2600 gas sensors (used in the upper row) and the TGS 2620 sensors (used in the lower row), this result indicates

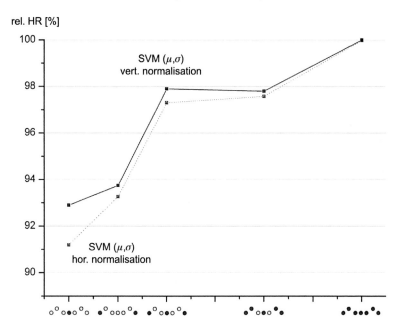

Figure 8.10: *Average relative classification performance obtained with different sensor combinations. The results were computed using the mean and standard deviation of each sector as the feature vectors.*

that a comparison of sensor data that were instantaneously sampled in a plane parallel and near to the floor is more important to detect proximity to a source of evaporating liquid ethanol than a comparison of sensor data sampled orthogonal to the floor. This is likely to be a consequence of the fact that ethanol, which is heavier than air, tends to stay near the floor.

8.7   Conclusions

This chapter was concerned with the task of gas source declaration. It introduced a classification method based on gas sensor readings only. In order to decide whether a gas source is in the direct vicinity, the robot collects gas sensor readings while it performs a rotation manoeuvre in front of a suspected object. The manoeuvre was chosen because it is easy to implement, requires little space and provides temporally as well as spatially sampled concentration data.

A total of 1056 declaration trials were carried out, including 528 trials at the minimal distance to the gas source (positive examples) and 528 trials with a varying larger distance of between 5 cm and 100 cm (negative examples). Based on a subset containing only those negative examples with a certain minimal distance from the gas source, four classifiers were trained and evaluated by cross-validation. For this purpose, a multilayer feedforward neural network (MLFF) and a support vector machine (SVM) were trained using either vertically or horizontally normalised feature vectors. The classification performance was then maximised by means of a grid search considering the number of hidden neurons in the case of the MLFF, and the kernel parameter and the penalty parameter in the case of the SVM.

With optimised learning parameters the support vector machine always yielded slightly better results compared to the artificial neural net that was used in this investigation. Thus, only the SVM classifiers were considered in order to determine the dependency of the cross-validation hit rate on the average distance of negative examples from the gas source. The results were then compared with the classification performance that could be achieved by selecting an optimal threshold value regarding the mean sensor signal during the rotation manoeuvre ("threshold classifier"). Here, the performance of the SVM classifier with optimised learning parameters was generally higher. While such a result would be expected in the case of vertical normalisation, which preserves information about the absolute intensity and the relative course of the sensor readings, it is more surprising in the case of horizontal normalisation, where the resulting feature vectors only contain information about the relative course of the sensor readings. According to expectation, the classification performance was always found to be higher when vertically normalised feature vectors were used instead of horizontally normalised ones.

The profile of the classification rate with respect to a decreasing average distance of negative examples from the gas source exhibits three distinguishable regions. For the largest and lowest average distances, an approximately linear descent was found with an approximately constant area when trials were considered that were recorded at a distance of 30 cm to 50 cm from the source (corresponding to an average distance of 60 cm to 72.5 cm). This could indicate a transition between domains where proximity to the gas source can be detected using different properties of the concentration field.

Table 8.4 itemises the maximal hit rate achieved with the SVM using vertical and horizontal normalisation considering two different data subsets with an average distance of 80 cm and 50.6 cm of the negative examples from the gas source. The corresponding rate of false positives and false negatives is also given in the table. This performance could be improved by combining single predictions obtained from data recorded at different positions. One possibility would be to repeat the rotation manoeuvre three times at a different orientation with respect to the source and to use the majority vote as the new prediction. Thus, a majority vote is guaranteed while

single estimates can be assumed to be independent due to the different direction of
the convective airflow at different orientations of the robot relative to the gas source.

Using the majority vote of three classifiers, the total rate of false positives FP^{tot}
and false negatives FN^{tot} as well as the total hit rate HR^{tot} can be computed from
the rate of false positives and false negatives of a single classifier (FP, FN) under
the assumption of independent estimates as

$$FP^{tot} = FP^3 + 3FP^2(1 - FP), \qquad (8.7)$$

$$FN^{tot} = FN^3 + 3FN^2(1 - FN), \qquad (8.8)$$

$$HR^{tot} = \frac{N_p(1 - FP^{tot}) + N_n(1 - FN^{tot})}{N_p + N_n}. \qquad (8.9)$$

Here, N_p and N_n are the number of positive and negative examples. The hit rate
that is expected for such a combined classifier is also itemised in Table 8.4.

It was mentioned that subsequent measurements carried out at different orientations of the robot relative to the gas source can be assumed to be independent
due to the different direction of the convective airflow. It is expected, on the other
hand, that the classification performance could be improved by using multiple series of measurements recorded at different positions (as one feature vector), because
in this way the difference between the sensor response at the downstream and upstream side of the gas source could also be used for classification. Ongoing work is
concerned with this issue. As a further extension of this research, the influence of
the maneouvre for data acquisition on the classification results is to be investigated,
and the capability of the trained classifiers to extrapolate on unknown situations
(different room, different gas source) should be tested in a validation experiment.

Future work will also investigate the question of what are the important features
required for classification (feature selection), the possibility to improve the classification performance by combining estimates from several local models, the prediction
capability of the distance to the source (regression) and the suitability of Bayesian
learning techniques.

Δd_{min}^{ns}	$\overline{\Delta d^{ns}}$	v-norm. HR(FP, FN)[%]	comb. v-norm. HR(FP, FN)[%]	h-norm. HR(FP, FN)[%]	comb. h-norm. HR(FP, FN)[%]
60 cm	80 cm	88.9 (9.6,12.7)	96.5 (2.6,4.4)	79.2 (22.7,19.0)	88.7 (13.1,9.5)
20 cm	50.6 cm	73.7 (23.7,29.0)	82.7 (14.2,20.4)	67.8 (32.3,32.1)	75.6 (24.6,24.3)

Table 8.4: *Comparison of the classification performance that could be achieved using a single SVM classifier with the performance that is expected for a combined
classifier, which uses the majority vote of three independent SVM classifiers. The
hit rate (HR), the rate of false positives (FP) and the rate of false negatives (FN)
is given obtained from vertically normalised feature vectors (v-norm.) as well as
horizontally normalised feature vectors (h-norm.).*

Chapter 9

Conclusions and Future Work

This thesis is concerned with two fundamental problems in the domain of gas sensing with mobile robots: gas distribution mapping and localisation of a static gas source.

The experiments presented were carried out with two different mobile robots corresponding to two possible application domains. The small scale system can be thought of as a study of a commonly available mobile robot that is equipped, or can be additionally supplied at low cost, with a gas-sensitive device. By contrast, the medium scale system can be regarded as a prototype of a dedicated gas-sensitive mobile robot that is intended to be used in larger environments or under difficult conditions (for example, in a rescue mission).

In an industrial or domestic indoor environment with moderate ventilation, wind speeds are typically encountered that cannot be measured with state-of-the-art anemometers and are generally hard to detect. With regard to real-world applications, the experimental environment was therefore not artificially ventilated in this work to produce a strong unidirectional airflow (uncontrolled environment). In contrast to previously suggested strategies for gas source localisation, which include periods of upwind movement, the methods suggested in this thesis do not rely on information about the air flow.

Apart from the limitations of current sensor technology, the main difficulty results from the spreading of gases under natural conditions. Diffusion plays only a minor role in the distribution of odourant molecules at room temperature. The dominant transport mechanisms are convection flow and turbulence. As a consequence, the resulting concentration field is patchy and chaotic, and the actual gas source is usually not located at the point of highest concentration.

Despite these characteristics of a turbulent gas distribution, a rough correlation between sensor response and proximity to a gas source can be obtained in simple scenarios with a constrained search space. It was found that response peaks of the gas sensors often correspond to the approximate location of a gas source in an uncontrolled corridor-like environment when the gas source localisation task is restricted to one dimension. Such a correlation, however, was only observed when the gas sensor measurements were performed while the robot was driven with a constant, sufficiently high speed ("constant-velocity sensing"). In this case, the sensor signal represents a spatially averaged concentration value, because integration over subsequent measurements is implicitly performed by the metal oxide sensors due to their long decay time. Thus, the many local concentration maxima around the gas source appear as a single peak, which is likely to be centred near to the source location, because of the higher density of local maxima in the vicinity of the gas source. This result demonstrates that the density of local maxima provides a feature that is better suited for gas source localisation than the maximum intensity of the instantaneous concentration field, which does not exhibit considerable variation with increasing distance to the gas source. It has to be mentioned, however, that response peaks of the gas sensors provide only a rough estimate of the gas source location, even if a suitable constant-velocity sensing strategy is applied.

In contrast to the experiments in a corridor-like environment, the location where maximum response peaks were obtained did not provide a satisfying estimate of the gas source location in the experiments in a two-dimensional scenario. Accordingly, a major part of this work is concerned with this task. A reasonable approach is to break down the full gas source localisation problem into three subtasks, which can be studied independently. First, the robot has to make contact with the target gas in order to start the search (gas finding). The gas finding problem is not addressed here. It amounts to a basic search task and the selection of a suitable threshold value. The second step is then to follow the cues determined from the sensed gas distribution in order to approach a gas source (gas source tracing). Finally, a method is required to establish that the source has been found in order to terminate the search (gas source declaration).

Two reactive gas source tracing strategies were suggested in Chapter 6 and Chapter 7 and evaluated in an uncontrolled indoor environment by statistical means. These strategies are based on a different view of the gas distribution. The idea of sensing the concentration gradient with a pair of spatially separated gas sensors (tropotaxis) was implemented in the manner of a Braitenberg vehicle, i.e., a control algorithm that applies a direct sensor-motor coupling was used. With uncrossed inhibitory sensor-motor connections the robot follows the local gradient towards the stronger signal. The resulting "exploration and hillclimbing" behaviour will be successful in the case of a smooth distribution with a maximum at the actual location of any gas source. In other words, a tropotaxis behaviour refers to the time-averaged gas distribution, which exhibits smooth gradients. Accordingly, a tropotaxis behaviour decreases the path length the robot needs to move to the source *on average* compared to random exploration. This result was found after a total of 36.5 hours of experiments with high statistical significance.

In contrast to the tropotaxis behaviour, the second gas source strategy suggested in Chapter 7 does not assume a smooth concentration field. Following the behaviour by which male silkworm moths localise a mate as a pheromone source, this strategy consists of a fixed motion pattern that is triggered by increased sensor readings. Since the motion pattern realises a local search for the next patch of gas, this biomimetic gas source tracing strategy rests on the assumption of a gas distribution that consists of comparatively isolated patches with high concentration of the analyte gas, surrounded by areas where the concentration is considerably lower. In the case of the moth, the motion pattern is oriented against the upwind direction. In order to avoid dependency on information about the air flow, a modified version of the moth strategy was suggested that uses a different mechanism to orient the motion pattern according to an estimate of the direction to the source based on gas sensor measurements only. An implementation of the proposed algorithm on a real robot was tested in an uncontrolled indoor environment and the results were evaluated by statistical means. It was found that the modified gas source tracing strategy decreases the average robot-to-source distance compared to random exploration. However, due to the relatively small number of experiments (10 hours in total) as well as the limited space, the obtained confidence level was not strong enough to justify a statement with high statistical significance.

Because the suggested gas source tracing strategies were implemented on different robots and tested in a different environment, it is not possible to compare the results directly. It is likely, however, that the biomimetic gas source tracing strategy is better suited when the distance to the source is large, because the assumption of smooth gradients is not at all fulfilled in the distal concentration field. In contrast, a tropotaxis behaviour might be superior, i.e., produce a shorter path, near to a gas source. Here, the assumption of smooth gradients is better fulfilled, especially considering the low pass filtered sensor signal obtained with the metal oxide sensors, because patches of gas occur more frequently. However, an experimental validation of this proposition, preferably in a larger area, remains to be done.

In order to provide a solution to the complete gas source localisation problem, a gas source tracing strategy needs to be extended by a gas source declaration mechanism. A method to classify a suspected object as being a gas source or not from a pattern in a series of spatially and temporally sampled concentration data is introduced in this work. The pattern was determined by applying machine learning techniques (artificial neural networks and support vector machines) to a set of more than 1000 training examples obtained in experiments carried out in an uncontrolled indoor environment. The results demonstrate the feasibility of the approach and show that high classification rates can be achieved using support vector machines. Consequently, a possible solution for the gas source localisation problem in an uncontrolled environment that is not corridor-like, and where the search space cannot be restricted to one dimension, would be a combination of the introduced gas source tracing strategies with the suggested gas source declaration mechanism. Either the gas source tracing behaviour that is better suited under the given conditions might be used or a combination of both together with a suitable switching mechanism. In order to accelerate the process of gas finding, an appropriate exploration behaviour might also be added. It will be a topic of future work to assemble a mobile system for gas source localisation based on these suggestions.

A third main issue that is addressed in this work is gas concentration mapping. With the introduced concentration mapping algorithm it is possible to create a two-dimensional gridmap representation from sparse point samples, which stores belief about the average relative concentration of a detected gas. In order to obtain complete concentration gridmaps, the path of the robot carrying the gas sensors should roughly cover the entire space, although a perfectly even coverage of the inspected area is not necessary. It is also advantageous to pass particular points from multiple directions to increase spatial accuracy. Furthermore, the current version of the mapping algorithm requires accurate location estimates. While an extended version might comprise a model to account for positional inaccuracies, a rough estimate of the location of the gas sensors will be required as well. However, if position estimates are available and the path of the robot covers the designated area, gas distribution mapping can always be used in addition to the previously mentioned gas source localisation strategies. Apart from providing a concentration map that might indicate contaminated areas, a combination with one of the suggested gas source localisation strategies could also make the gas source declaration process more reliable, because the maximum cell in a concentration gridmap is often a good estimate of the location of the gas source. While gas source tracing and gas distribution mapping were basically studied independently in this thesis, future work should address suitable strategies for a system that performs both tasks simultaneously, i.e., strategies for gas source tracing, which generate search paths that comply with the above mentioned requirements for gas distribution mapping.

A gas source localisation strategy, which combines gas source tracing with a gas source declaration mechanism as suggested, assumes that the source appears as an obstacle to the robot. In cases where the source cannot be sensed as a single, isolated obstacle, however, this strategy has no means to determine possible candidates for the gas source declaration method. In order to overcome this limitation, there are two conceivable extensions. First, the gas source tracing algorithm could be augmented by a mechanism that indicates possible candidates for gas source declaration. In the case of the mentioned biomimetic gas source tracing strategy, for example, such a mechanism could be based on the rate at which the fixed motion pattern is triggered. It was found in the experiments that vicinity to the gas source often corresponds to a continued period of high triggering frequency. Thus, a high triggering frequency could be used to start a declaration method to determine the certainty that a gas source has been found.

An alternative method for gas source localisation in the case that the source does not take the form of an obstacle, would be a gas source tracing strategy that includes the declaration step. An example is the gas-sensitive Braitenberg vehicle with crossed, inhibitory connections that was also experimentally tested and statistically evaluated in Chapter 6. The resulting behaviour can be described as "exploration and concentration peak avoidance". With crossed sensor-motor connections the robot evades local concentration maxima while a simple sort of exploration behaviour is carried out because of the inhibitory character of the connections. Due to the fact that local concentration maxima occur more frequently in the vicinity of the gas source, the path of the robot covers the available area less densely near the actual location of the source. A picture of the path of the robot (more precisely: the path of the gas sensors) thus reveals the location of a gas source by a comparatively unexplored area. This result again emphasises the important insight that the density of local maxima is better suited as a feature for gas source localisation than global maxima of the instantaneous concentration field.

In order to exploit this feature, however, an implementation as a Braitenberg vehicle is certainly not optimal, especially concerning time consumption and the non-linearity of the sensors. There is no mechanism that causes a preference for even exploration and thus much time is wasted until all the possible locations other than the gas source can be excluded. Although the search time scales down with the number of robots utilised, more efficient strategies that search for high densities of local concentration maxima in order to locate a gas source are clearly demanded, considering that the time consumption increases with the size of the area observed. Such strategies will be also part of the future work.

Of course, such strategies could be combined with an additional gas source declaration mechanism to determine the certainty that the unexplored area contains a gas source. Furthermore, a combination with gas distribution mapping would also be desirable. In the case of the Braitenberg vehicle with uncrossed sensor-motor connections ("exploration and concentration peak avoidance") reasonable concentration maps could be created from the recorded data because the path of the robot complied with the requirements for gas distribution mapping on average. Consequently, future research should also address gas source localisation strategies based on a search for high densities of local concentration maxima with regard to their suitability for gas distribution mapping.

Within the timeframe of this thesis, several methods for gas source localisation and gas distribution mapping have been investigated in an uncontrolled indoor environment. Suggestions for future work regarding the individual methods are given at the end of the corresponding chapters. A natural extension of this work is now to combine these techniques and to investigate the performance on a mobile robot under different conditions. In particular, larger indoor environments and also outdoor conditions will have to be considered. An investigation of a complete mobile system for gas source localisation comprising several of the techniques suggested in this thesis will also have to address the question of how the different modules can be combined most efficiently with respect to a given task. The next step will then be to assemble prototype robots for selected industrially relevant applications and to carry out long running tests with these systems. Finally, the prospect of the various possible applications for gas-sensitive mobile robots might boost the development of gas sensors especially designed for use on a mobile robot. And this again will result in new challenges in this fascinating research area.

Appendix A

Software Architecture DDFLat

This chapter introduces the DDFLat framework, which was used to develop all the programs for this thesis. The framework was primarily intended to support developing robot control software by providing an efficient, modular structure that allows to adjust the real-time interaction of the program modules and includes a demonstrative way to visualise how an application works. Because most of these advantages are also beneficial to solve programming problems in other domains, it was found to be advantageous to use the DDFLat framework for all the programs that contributed to this work.

A.1 Towards "Good" Robot Control Software

In order to understand the requirements for developing robot control software, it is useful to point out what demands "good" software in this domain usually has to fulfil. A mobile robot interacts with the environment by processing information from its sensors to select appropriate actions. (This characteristic is so fundamental that it was used by Brady to define the field of robotics as the "intelligent connection of perception to action" [Bra85].) Interacting with a dynamic world requires the robot control program to obey timing constraints. First, this implies a need for *efficiency*. Therefore a framework for robot control applications should not introduce restrictions that prevent the developer from using all of the available resources. Furthermore it is desirable to provide a possibility to *adjust the real-time behaviour* of a program easily. Consider, for example, a robot that performs vision-based localisation and that the actual implementation needs 200 ms to calculate a new position estimate. Although the robot might get images from the camera with a rate of 25 Hz, it would of course not be able to provide position estimates with a rate higher than 5 Hz. Let us assume that this robot is now extended by a second subsystem that uses laser range finder readings to provide alternative position estimates. For the sake of simplicity, it might also take 200 ms to calculate these estimates, which might be again too slow to utilise all the readings provided with a rate of 40 Hz. Because the combination of the localisation methods increases the reliability of the overall position estimate, it is desirable to use both of them. If both subsystems would run in their own thread of execution, the average rate of position estimates would be 2.5 Hz. Depending on the situation considered it can, however, be advantageous to prioritise one of the methods by tuning the assigned execution time. Imagine that the second localisation method produces a very accurate estimate but is susceptible to perceptual aliasing[1] and relies on a particular model of the environment (for example by assuming wall-like structures that are internally represented by straight lines). In this case it is reasonable to perform the accurate localisation method more often, for example with a rate of 4 Hz, while the vision based method is used rarely (1 Hz) to solve ambiguous situations due to perceptual aliasing. In other situations it can be necessary to change the assigned execution time dynamically. If the robot for example leaves a building and thus the applied model of the environment is no longer valid, the corresponding localisation method becomes less useful. Now the vision-based localisation strategy should be executed more often, while the alternative method should be used just rarely to check whether the underlying model becomes applicable again.

Besides the demand for efficiency and a tunable real-time behaviour of subsystems, there are further requirements that arise as a consequence of the way robot control software is tested and debugged. For the purpose of testing and debugging it is necessary to test a running program in a realistic dynamic environment, either a simulated one or – preferably – an authentic real world scenario. Conventional debuggers that allow to halt program execution and to scan the current values of the relevant variables are only applicable in the case of a simulation that allows one to halt the dynamics of the environment. In the case of testing in a real world environment, this is of course not possible. Here, the value of all possibly relevant variables has to be logged and evaluated afterwards. Hence, an obvious but often disregarded rule is to keep the source code of the application to be tested clearly separated from the code that is needed for simulation or for logging the data. Thus, a framework for robot control software should support the *decoupling of the main application from components needed for simulation or data logging.*

Writing robot control software from the scratch is an intricate task. However, different applications usually share a considerable proportion of components such as hardware drivers, data preprocessing algorithms or modules for localisation, path planning, etc. A reasonable framework should therefore strongly *support the reusability of particular components.* This again requires the decoupling of the functional building blocks of a program. If these building blocks are expected to be shared among several developers, the aspect of acceptance is important. The function of particular components, as well as the way these components cooperate, should be easy to understand and to communicate. In this respect a framework may assist

---

[1]Perceptual aliasing refers to the situation where several places are perceptually similar enough to be confused by the robot [Duc00].

the development process by *providing a consistent structure and a common vocabulary* to facilitate documentation and thus communication among developers. Note that this implies the need for a plain framework. Otherwise the benefit of an improved communication could be easily outbalanced by the effort to understand the framework itself. Additionally, a consistent design that supports loosely coupled building blocks aids also the development of new components, because each component can be treated as an independent task by another developer. Likewise testing of alternative implementations of particular components is supported in this manner.

The requirements itemised in this section apply to problems in other domains, too. Consequently, the software framework to meet these requirements, which is introduced in the following sections, is not restricted to the field of robot control. All of the programs needed for this thesis were implemented using this framework, including the software to control external measurements to perform data evaluation, and to simulate sensor input. An example of a robot control application is given in the following sections to illustrate the DDFLat framework. As a further example of an application in a different domain, the software, which is used for the absolute positioning system W-CAPS, is presented in Appendix B.

A.2 Basics of the DDFLat Framework

In order to meet the demands specified in the previous section, the DDFLat framework (**D**ynamic **D**ata **F**low with **Lat**ency) was introduced. The main idea of this framework is to map the functional units of an application to objects and to model the cooperation between these objects by dynamically configurable data flow chains. A data flow chain is represented by a cascade of connected objects, which are updated downstream if a certain object changes its internal state. The possibility to adjust the timing of the processes involved is provided by a latency period that can be assigned to each object, meaning that the object cannot trigger an update cascade before this period has elapsed. Thus, a maximum update frequency can be specified for each object, and consequently also for each part of the data flow chain. Rather than presupposing well-defined and fixed response times, the DDFLat framework provides a developer with the possibility to tune the timing of the whole system such that it meets the given timing constraints acceptably on average. This means that soft real-time systems were considered[2]. A latency-based mechanism was chosen to avoid a restriction to real-time operating systems (RTOS) that ensure a certain capability within a specified time constraint. All the applications that

---

[2]In this thesis the term *real-time system* is used to refer to software systems where the correct functioning of the system depends on the results produced by the system and the time at which these results are produced. Hard and soft real-time systems are distinguished by the consequence it has if the system fails to meet the timing constraints. In case of *hard real-time systems* an incorrect operation results from timing errors while for *soft real-time systems* the operation is just degraded [Som01].

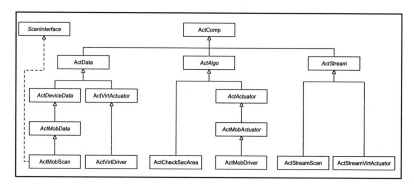

Figure A.1: *Hierarchy of the classes needed to create the security data chain for the example application that is developed in this section to illustrate the design of the DDFLat framework. The functionality is mainly provided by the abstract base classes* ActComp *and* ActAlgo.

contributed to this work were implemented on a Linux system using the DDFLat framework.

In the following, the functionality of the DDFLat framework is explained by gradually constructing an application that can be used for gas concentration measurements in a corridor as described in Chapter 4. Here, the robot should perform concentration measurements while driving up and down along a corridor following its middle. Some of the classes that were needed for this application are pictured in Fig. A.1, which shows a cutout of the complete class hierarchy using the Unified Modeling Language notation (UML 1.0) for class diagrams [RJB97]. All classes inherit the common base ActComp[3]. This abstract class offers methods to create and trigger data flow chains and implements the automatic update of objects downstream. In order to start with the implementation we will assume that all the classes needed are available for the moment. How further classes can be added is explained in Section A.2.4.

Being responsible roboticists, we have to think of course about safety first. Thus, we start by creating a data chain that generates stop commands when the robot

---

[3]All the class names start with the prefix "Act" for historical reasons. Instances from these classes are "active components" or "active objects" in the sense of an object for which "modification of an attribute can cause certain activities" [Som01]. Consequently the framework was called AORC (Active Object Robot Control) first. To avoid confusion with the design pattern "Active Object" [SSRB00], which is intended for concurrent and networked applications to decouple method invocation and method execution, the denotation was changed to the more accurate term DDFLat later on.

approaches an obstacle. The stop command is represented by an instance of the class `ActVirtDriver`, which is a data structure that holds a driving command (for example two values that determine a translational and a rotational speed) and a priority. It provides three groups of methods: the methods that enable insertion in a data flow chain (inherited from `ActComp`), a second one that allows to read and set the priority of the command (inherited from `ActVirtActuator`) and a third group that provides access to the internal data structure. Every object that represents a data structure inherits `ActData`. Though it is possible to build a chain solely consisting of connected data objects, this is not very meaningful because data objects are not intended to change each others internal state.

DDFLat offers the base class `ActAlgo` to perform operations on data objects. Derived classes have to implement a method, which can use the data objects connected upstream as input (the data source) and the data objects connected downstream to output the results computed (the data sink). For the example considered an algorithm is needed, which checks a certain security clearance and produces a stop command if the robot runs the risk of crashing into an obstacle. To achieve this, the class `ActCheckSecArea` can be used. By comparing a set of range data (provided by a connected source) with internally stored lower bounds for each distance measurement, instances of this class watch a "security area" and produce a high priority stop command in case of an intrusion to this area (see Fig. A.7).

To supply distance measurements the robot is supposed to be equipped with a laser range finder, which is represented by an instance of the device class `ActMobScan`. Device classes represent the data provided by the device. Hence, they inherit `ActData`. In addition they encapsulate the communication with the device by implementing the low level protocol for setting the parameters of the device, polling it, etc. Thus, device objects act as hardware drivers and provide the sensor readings in a data structure that can be used by connected algorithm objects. The example given in this section corresponds to a real implementation on a robot that utilises a CORBA-based communication model to distribute sensor readings. Therefore, devices that were installed on this robot inherit a common base class (`ActMobData`[4]), which provides methods to use this intermediate layer. The class `ActMobData` inherits the base class for devices `ActDeviceData` that contains methods to poll for new data and to check whether the device is already initialised. In order to guarantee that algorithm objects can access sensor data in the same way on different platforms, device classes that correspond to the same type of sensor inherit a common interface class. A laser range finder class inherits the class `ScanInterface`, for example, which declares methods to access particular range values and also methods to determine the number of readings per scan, the angle covered by a full scan, etc.

---

[4]The class was called `ActMobData` because the CORBA functionality is used via a library called Mobility provided by the manufacturer of the robot.

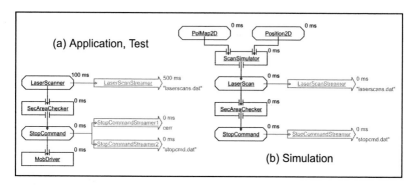

Figure A.2: *DDFLat diagrams of two versions of a security chain.*

Now the security chain is almost complete. For an application on a real robot it just remains to add a component that actually executes driving commands. Any actuator is represented by an object of type `ActActuator`, which executes connected commands. While device data objects are used at the root of chains, actuator objects terminate branches of them. To close the security chain in our example we can use the class `ActMobDriver`, which performs driving on the chosen platform. Additionally, this class can be applied to realise acceleration limits or to verify driving commands before execution, thus preventing implementation errors from causing severe damages to the robot or its surroundings.

A.2.1   DDFLat Diagrams

The complete security chain is shown in Fig. A.2. To introduce the elements of such diagrams the version (a) on the left side is discussed first. For each object the assigned latency period is given by a value that is indicated near the upper right corner of the box around the objects name. Data objects are displayed by boxes with clipped corners while algorithms are represented by unclipped boxes with in- and outlets for each data type that is required by this algorithm. Actuator objects at the end of a chain do not have any outlet. Connections are displayed by arrows pointing in the direction of the data flow. Illustrations such as those shown in Fig. A.2 are referred to as DDFLat diagrams in the following discussion.

The security chain described so far is composed of three parts: a device object that represents sensor input (`LaserScanner`[5]), an actuator (`MobDriver`) and the remaining part that implements the main functionality (`SecAreaChecker` and `StopCommand`). A fourth group also indicated in Fig. A.2 (a) is constituted by

---

[5]Following the UML notation, class instances are indicated by an underline.

streaming objects of type `ActStreamScan` and `ActStreamVirtDriver` (both derived from `ActStream`). These objects are intended for testing and debugging DDFLat applications. If such an object is connected to a compatible data object, it writes the current data values to a selected stream, while keeping the specified latency condition. A stream represents an output device like a console or a file. In DDFLat diagrams streaming objects are represented by lightly drawn arrow-shaped boxes containing the identifier of the object. The latency period is indicated at the upper right and the stream is specified near the lower right of the arrow.

DDFLat diagrams visualise the process flow that is realised by the software implementation. The complete security chain in Fig. A.2 (a) shows a process that involves: reading laser range scans (with a maximum frequency of 10 Hz), writing some of these scans to the file "laserscans.dat" (every fifth scan if possible, with a frequency of ≤ 2 Hz), and stopping the robot immediately if an obstacle is detected inside the security area (see Fig. A.7) kept under surveillance. Each stop command produced is written to the console and to the file "stopcmd.dat".

The objects that implement the core functionality are used in exactly the same way in a simulated environment. As can be seen in Fig. A.2 (b), the subchain containing `SecAreaChecker` and `StopCommand` remains unchanged. In contrast to the actual implementation on a robot, the laser range scans are not provided by a device object. Instead the algorithm `ScanSimulator` computes range values according to a virtual laser scanner, using the position of the robot (`Position2D`) and a polygonal representation of the environment (`PolMap2D`). These objects are instances of the classes `ActMap2D`, `ActPosition2D`, `ActScan` and `ActScanSimulator`, which are not shown in Fig. A.1. The latter one inherits `ActAlgo`, while the former three are of type `ActData`. `ActScan` also inherits `ScanInterface`. With the security chain version shown in Fig. A.2 (b), laser range scans are simulated if a new robot position or a new representation of the room is given. Each new scan is written to the file "laserscans.dat" and triggers the algorithm to check the security area. Finally the stop commands produced are written to the file "stopcmd.dat".

A.2.2 Establishing Data Flow Chains

The source code to assemble the security chain that is shown in Fig. A.2 (a) is partly quoted in Listing A.1. Here, the basic structure of a DDFLat program can be seen. At the beginning, all of the objects needed are created (see lines 6 – 14). For each instance, three properties have to be specified: the latency period (given in milliseconds), an identifier and a reference timestamp. Each object stores a momentary timestamp that is used to check the latency condition. This timestamp might correspond to the current time at which data were acquired or an algorithm was executed, but it can also be the timestamp of the object that triggered the cascade and thus led to an update of the instance considered. This point is addressed in detail below. To check whether the latency for an object has expired, the momentary timestamp is compared with an internally stored update timestamp that indicates the time at which the object can be updated again at the earliest. The update timestamp is increased by the latency period each time an object is updated. To synchronise the data flow, a common starting point for this calculation is necessary. Therefore, a reference timestamp (see line 2 – 3) must be specified in the third parameter of the constructor.

In a next step, the objects created are initialised using the interface of the derived class. As an example, a rectangular security area for the robot to watch is specified in line 17 for the algorithm SecAreaChecker. Finally the objects are connected with a series of subscribe() calls (see lines 20 – 25). This method tries to establish a directional connection from the object for which the method is called to the argument passed. Thus, the arrow operator used to call the method indicates the direction of the connection.

```
   // Init Reference Timestamp
 2 TimeStamp tsRef;
   tsRef.set(); // set reference time = current time
 4
   // Create DDFLat Objects
 6 ActComp* LaserScanner =
     new ActMobScan(100.0,"SICK Laser Scanner",tsRef);
 8 ActComp* LaserScanStreamer =
     new ActStreamScan(500.0,"Laser Scan File Streamer",tsRef);
10 ActComp* SecAreaChecker =
     new ActCheckSecArea(0.0,"Security Area Checker",tsRef);
12    // ... and so on ...
   ActComp* MobDriver =
14    new ActMobDriver(0.0,"Mobility Driver",tsRef);

16 // Init DDFLat Objects
   ((ActCheckSecArea*)LaserScanStreamer)->setRectSecurityArea(0.5,0.25);
18
   // Connect DDFLat Objects
20 LaserScanner->subscribe(SecAreaChecker);
   LaserScanner->subscribe(LaserScanStreamer);
22 SecAreaChecker->subscribe(StopCommand);
   StopCommand->subscribe(StopCommandStreamer1);
24 StopCommand->subscribe(StopCommandStreamer2);
   StopCommand->subscribe(MobDriver);
26
   // Use of DDFLat Data Flow Chains
28 for(;;) {
     LaserScanner->touch();
30   usleep(SOME_TIME);
   }
```

Listing A.1: *Assembly and use of a DDFLat Data Flow Chain*

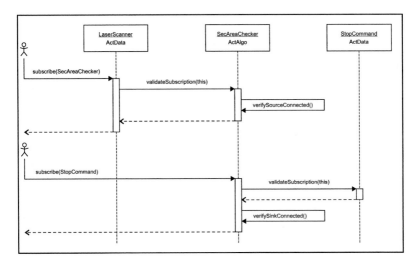

Figure A.3: *Sequence diagram of the subscription process.*

The Subscription Process

What exactly happens during the subscription process (which is the process of establishing connections between the objects) can be seen in Fig. A.3. The sequence diagram [RJB97] shown indicates the interaction of the objects caused by the method calls in line 20 and line 22 of Listing A.1. `ActComp` implements the basic procedure: First, the instance for which `subscribe()` was called (the source) adds the pointer passed (the sink) to an internal list (the sink list). Then, it uses the same pointer to call the virtual method `validateSubscription()` and passes a pointer to itself as an argument. This is a mirrored version of the initial `subscribe()` call, which causes the pointer passed to be added to the source list of the sink. Such a base class implementation ensures that the internal lists contain the pointers to all of the connected sinks and sources. Both `subscribe()` and `validateSubscription()` check the pointer passed and throw an exception in the case of unsuitable objects (to avoid multiple subscription of the same instance for example). Because an entry in the sink list is erased if the subsequent `validateSubscription()` call causes an exception, it is ensured that each pointer in a sink list corresponds to an equal pointer in the source list of a connected object.

The class `ActData` does not change the subscription process already described for the implementation of `ActComp`. By contrast, the class `ActAlgo` extends the subscription process with methods to check the suitability of the data objects referenced in the sink and source list. As indicated in Fig. A.3, the implemen-

tation appends a call for the virtual method `verifySourceConnected()` in
`validateSubscription()` and a call for `verifySinkConnected()` in sub-
`scribe()`, respectively. In the overridden version of these virtual methods, derived
algorithm classes have to determine whether it is possible to use the algorithm with
the objects connected. This includes ascertaining the validity of the listed pointers and might also comprise checking the correct initialisation of the referenced
instances. The former task is carried out by examining the type of the objects.
After `subscribe()` is called for instances of the class `ActCheckSecArea`, for
example, `verifySinkConnected()` tries to cast the pointer passed to a pointer
on `ActVirtDriver`. Thus, it is guaranteed that the implementation can use the
given pointer to store stop commands. In the same way, source pointers are casted
tentatively to `ScanInterface` in `verifySourceConnected()`. By ensuring
that the pointers can be used in the intended way during the subscription process,
the actual implementation of an algorithm can efficiently access the required data
structures afterwards. Thus, connected data objects can also be seen as dynamically
linked attributes of the algorithm class. To ensure proper operation, it is sometimes
necessary to check the initialisation of connected objects, too. In the example of
the algorithm `ActCheckSecArea`, the security area is defined either by specifying
a geometric shape like a semicircle or rectangle, or by providing a list of minimum
distances that constitute an arbitrary polygonal shape. In the latter case, it has
to be ensured that the size of this list matches the number of range values per
scan provided by the connected instance of `ActMobScan`. In short, the methods
`verifySourceConnected()` and `verifySinkConnected()` have to certify
that the algorithm is usable within the data flow chain it is put in. For this purpose, internal flags are set to indicate whether the required data sources and sinks
are already connected. These flags are used by the method `usable()` provided
by `ActAlgo`, to reply to queries on whether an algorithm is available. During an
update cascade, algorithm objects are only invoked if `usable()` confirms a proper
setup. Thereafter the actual implementation uses the linked instances without further examination. The mechanism described above is also invoked if the registration
of objects is erased. Thus, at any time, algorithm objects are guaranteed to be invoked only if they are properly integrated into the current data flow chain.

A.2.3   Update Cascades

To understand the basics of the DDFLat framework, it just remains to discuss the
data flow process. An update cascade can be triggered with the method `touch()`,
which declares that the state of the object for which it was called has changed
and sets the momentary timestamp of the object to the current time[6]. If called
for device objects (as in line 29 of Listing A.1) the method `touch()` also includes
initial polling of the corresponding device. In the case of new data being available,

---

[6]Compare the UNIX command "touch" that sets the modify date of existing files to the present.

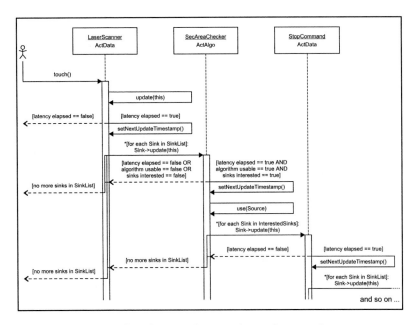

Figure A.4: *Sequence diagram of an update cascade.*

these data were read out and a state change is declared. If the underlying system provides events, the `touch()` call can be linked to the event that is generated if new sensor data are available. Otherwise an event-driven system has to be simulated with a loop like the one in line 28 − 31 of Listing A.1.

Referring to the security chain example once more, the effects triggered by a `touch()` call are shown in the sequence diagram in Fig. A.4. Updates of DDFLat objects are handed down by repeated calls of the method `update()`, which first checks if the object is allowed to be updated and − if so − informs all of the connected sinks by calling `update()` for each one of them. By supplying a pointer to the invoking source as an argument, the sink gets the opportunity to refer to the particular source that actually triggered its update. This is important mainly in the case of algorithm objects with several sources. To start an update cascade, the method `touch()` generates an `update()` call to itself. The `update()` method is implemented differently for data and algorithm objects. If called for an object of type `ActData`, only the latency condition has to be checked. If the latency period has not yet elapsed, the method simply returns. Otherwise the update timestamp is set by the method `setNextUpdateTimestamp()`.

The time when the next update can be performed is calculated by successively adding the latency period to the stored reference . Because a future timestamp (compared to the momentary timestamp) is always chosen, accumulation of updates is avoided during a period of blocked activation. Consider, for example, an algorithm with a latency period of one second that depends on sensor data it receives through a wireless network. Assume also that the next update timestamp would be determined by adding the latency period without ensuring future timestamps. If, for any reason, a problem with the wireless connection occurs, putting the transmission out of operation for 30 seconds, then the algorithm would be executed at least the next 30 times in a row the next time an update cascade is triggered. Although the chosen update frequency could be achieved more closely on average in this way, checking for future timestamps is performed to ensure a steady execution flow. Thus the situation is avoided where temporary disturbances can cause the timing to collapse for a period of time after the disturbance occured.

Objects for which the latency period has elapsed call `update()` for each registered sink after the update timestamp has been updated. For the example shown in Fig. A.2 (a), this means that <u>LaserScanner</u> informs both <u>LaserScanStreamer</u> and <u>SecAreaChecker</u>. By contrast to data objects that just check the latency condition, instances of type `ActAlgo` additionally try to determine whether it makes sense to perform the algorithm, which is encapsulated in the `use()` method. Therefore, the virtual method `usable()` is called first to query whether the algorithm is properly set up and the objects needed are connected. Next, a list of interested sinks is determined from the list of registered sinks. An interested sink is an object that is able to pass down the focus of control in an update cascade because its latency time has elapsed. Consequently the `update()` method has to be called for interested sinks only. To avoid computationally expensive calculations as much as possible, the actual algorithm is just performed if all of the conditions mentioned are fulfilled. Similarly, instances of `ActStream` are checked to see whether the encapsulated streaming process is usable, which might not be true because no stream has been specified yet. However, because there can be no sinks connected, checking for interested sinks is not necessary for streaming objects.

To trigger an update cascade, a timestamp can be specified in the initial `touch()` call, which is then assigned to all of the updated objects in the data flow chain. If used without an argument, the momentary timestamp is set to the current time before it is compared with the update timestamp to determine whether the latency period has elapsed. This is usually not desired in robot control applications. Here, the results determined from a certain package of sensor data should rather be assigned with the time when the sensor data were collected, not with the time when the object that represents the result was updated. To facilitate triggering of update cascades with the timestamp of the sensor data, the method `poll()` is provided for device data. It polls for new sensor data and, if new data are available, it performs a `touch(ts)` call with the timestamp `ts` of the data package.

A.2.4 Adding DDFLat Classes

This section describes how new classes can be added to implement a desired behaviour. The data flow chain discussed so far (see Fig. A.2) is responsible for stopping the robot in front of obstacles, which is of course only reasonable if the application generates positive driving commands as well. We will now continue with a data flow chain that implements a certain driving behaviour. Recall that the intention is to develop a program that performs gas concentration measurements while driving the robot up and down along a corridor following its middle. Therefore, an algorithm is needed to generate driving commands that try to align the robot with the corridor and to maximise the distance to the walls at the same time. Such an algorithm needs an idea of the corridor's geometry. For this purpose we will use features of the environment, which are extracted from the distance measurements provided by each laser range scan. Successive points that form straight lines in the scanned plane were combined to form segments, and afterwards the main axes of the environment are determined from the directions of these segments. Thus, a polygonal environment is assumed. One possibility would be to combine both the part that infers the geometric information and the part that calculates the actual driving commands in the same algorithm. But it is much better to separate the main steps of the calculation into different objects for two reasons: first, the algorithm that determines driving commands is decoupled from the process that calculates geometric information. Because the functional parts of the preprocessing chain can be easily exchanged, this facilitates testing of different preprocessing algorithms and makes it easier to switch to other methods that provide the same information later on (such methods might become feasible because of the addition of new sensors or further development of algorithms). Second, other algorithms can access intermediate results and do not have to perform the same calculations again. In the application we are currently developing, for instance, the main axes of the environment are used to align the robot with the corridor and also to correct for odometry errors (see the DDFLat diagram in Fig. A.5, which is explained in detail below).

Starting with the algorithm that determines the segments in a laser range scan, the first question to ask is about the data type needed as in- and output. In the case considered, the input is of type `ScanInterface` and the output is a data type that stores the segments calculated. To represent segments it is sufficient to store the coordinates of two points. But because subsequent algorithms might benefit from knowing which measurements contribute to a particular segment, an alternative representation was chosen that stores also the index of the first and the last of these measurements (`ActIndSegments`). For the calculation of the main axes, however, just the length and the direction of the segments is needed.

Implementing new data classes is simple. There is nothing more to do than to derive the new class from the common DDFLat base `ActComp` and to implement specific methods to access the data structure encapsulated. If legacy code is available, these specific methods can be inserted by inheritance from the existing class.

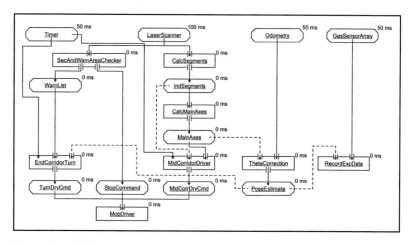

Figure A.5: *DDFLat diagram of an application that performs gas concentration measurements while driving the robot up and down a corridor following its middle.*

There is a little bit more to do to implement new algorithm classes. First, an algorithm has to verify the connected data sources and sinks. Therefore, the virtual methods verifySourceConnected() and verifySinkConnected() have to be overridden. These methods are called after a new sink or source is tentatively added to the internal sink/source list. Each algorithm has to check for the correct data type of the objects registered. Additionally, the subscription process often requires checking of consistency among the data objects connected, and among the data objects and the parameters of the algorithm. In short, the verify[Source-|Sink]Connected() methods have to ensure that the actual algorithm can be executed as intended during update cascades.

If a proper execution is possible, this has to be asserted by setting certain flags. Two such flags are provided by the base class ActAlgo to indicate whether all sinks and sources needed are available. These flags are used by the base class implementation of the method usable(), which is called within update cascades before the actual algorithm is executed, to determine, whether the algorithm can be utilised. If further consistency conditions have to be fulfilled, compliance with these conditions has to be checked in an overridden version of usable().

After the algorithm is properly integrated into the data flow chain, the actual implementation has to be realised by overriding the pure virtual method use(). Within this method the algorithm can refer safely to the connected data sources and sinks using pointers, which enables an efficient implementation. This direct access

- Assure proper integration into data flow chains by overriding `verify-SourceConnected()` and `verifySinkConnected()`.

- Overwrite `usable()` if additional conditions have to be assured besides that all the sources and sinks are connected properly.

- Implement the actual algorithm by overriding `use()`.

Figure A.6: *The 3 steps to be performed in order to add new DDFLat classes.*

to data objects is indicated in DDFLat diagrams by the fact that the connecting arrows start and end inside the boxes, which represent objects.

The method `use()` contains the complete implementation needed to perform the desired behaviour. Thus, the situation is avoided where the functional parts of the program are spread over the whole source code. This not only facilitates concentration on the functional part of a program, but also simplifies delegation of programming tasks to individual developers because they do not have to bother about the integration into the whole application. In addition, the fact that the functional part can be easily located might also attract other developers to integrate the implementation of the algorithm into their framework.

The steps that have to be performed in order to add new DDFLat classes are summarised in Fig. A.6.

The Complete Application

The whole application (see Fig. A.5) consists of three chains that produce driving commands, one chain that provides position estimates and another chain that performs the gas sensor measurements and writes them to a file together with time and position information. The parts the application is composed of are explained briefly below.

The chain that generates the commands, which drive the robot aligned with the walls in the middle of the corridor, is triggered by new laser range data (with a maximum frequency of 10 Hz). First, segments are extracted from a set of laser range values by regression. Second, the dominant orientation of these segments is determined. The unoriented angle of each segment is sorted into a fuzzy histogram [ADDK99], weighted by the length of the segment. Then, the average angle of the peaks in this histogram is calculated, which gives the orientation of the main axes in the environment.

The third algorithm in this chain uses both the set of main axes and also the set of segments the main axes were calculated from. To avoid execution of this

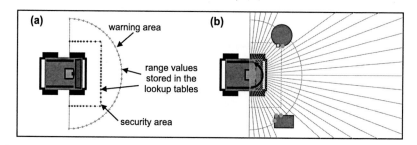

Figure A.7: *(a) Definition of warn and security area, (b) an example situation where the warning area as well as the security area is violated.*

algorithm immediately after a new set of segments becomes available, the object that encapsulates the segments subscribes to the algorithm <u>MidCorridorDriver</u> using the method `subscribePassive()` instead of `subscribe()`. Thus, the algorithm can access the connected data source but is not triggered by it.

Using the previously extracted features of the environment, the algorithm <u>Mid-CorridorDriver</u> produces driving commands aiming to align the robot with the main direction of the corridor, while trying to maximise the distance to the segments parallel to this main direction at the same time. The driving commands generated are assigned with a low priority. They are executed – if no commands with higher priority are queued – until another command is generated. For this reason, it is necessary to enable the algorithm to update the driving command frequently. Therefore, it can be triggered not only by new laser range scans, but also by objects of type `ActTimer`, which are special data objects that do not encapsulate an internal data structure. In this way, the ability to trigger execution of an algorithm directly regardless of whether source objects have been changed or not is provided. Such a timer can also be used to synchronise the execution of algorithms, as it is done in this application with the algorithm, which generates the command that drives the robot in the middle of the corridor, and the one that performs a 180° turn at the end of the corridor.

The leftmost chain in Fig. A.5, which is responsible for turning the robot on the spot at the end of the corridor, can also be triggered by new laser range scans. Instead of the algorithm that checks only a security area, as in the example shown in Fig. A.2, an extended version that monitors two areas is applied. These areas are represented with a lookup table that stores a minimum length for each ray of the connected scan, specified either by defining the dimensions of a geometrical shape or by itemising all the values of the lookup table. Fig. A.7 (a) shows an example of a rectangular security area and a semi-circular warning area, defined for an assumed laser range scan with 37 rays. The values stored in the lookup tables are indicated by small dots. In the lower part of Fig. A.7 (b), a situation is shown where a rectangular obstacle lies inside the warning area (the range measurements that are too low are indicated by small circles). If at least one of the distances measured is lower than the corresponding value in the lookup table, a warn list is generated containing all distances that are too low together with the corresponding angles. This information can be utilised by connected algorithms to avoid obstacles. Here, it is used by the EndCorridorTurn algorithm to determine whether the robot reached the end of the corridor. If so, driving commands are generated to perform rotation on the spot. To detect when to stop rotation, an estimate of the current orientation provided by corrected odometry information is used (the corresponding data flow chain is explained below). Because a medium priority is assigned to the driving commands generated by the EndCorridorTurn algorithm, the execution of the commands from the MidCorridorDriver algorithm is suspended temporarily until the robot is reversed, which is indicated by generating a terminating driving command with the lowest priority (IGNORE).

Stop commands (with the highest priority IMMEDIATE) are produced if an object enters the security area. The range measurements are checked in counterclockwise order. If a violation of the security area is detected – as indicated by the small square in the upper part of Fig. A.7 (b) – the algorithm SecAndWarnAreaChecker produces a stop command and returns immediately. This might result in an incomplete warn list, but guarantees that the robot is stopped as fast as possible.

With the three data flow chains already discussed, the robot is driven up and down the corridor. To simultaneously record the data collected with the electronic nose, another chain is used (the rightmost one in Fig. A.5), which stores the gas sensor readings as well as the time and position at which the corresponding values are received. To correct for odometry errors, yet another data flow chain is arranged. With the used robot, the measured orientation in particular tends to differ clearly from the true orientation after just a few meters of travel. To correct for this distortion, it turned out to be sufficient to combine the information about the differential rotations of the robot, obtained from odometry and from tracking the orientation of the main axes. This is performed by the algorithm ThetaCorrection, which determines how much the orientation of the current main axes differs from a set of reference angles stored from previously found main axes. Thus, an estimate of the differential rotation during the last timestep can be calculated, which is then combined with the corresponding odometry value in order to calculate the final estimate.

Using the corrected differential rotation, a sufficiently accurate position estimate could be obtained for the purpose of the intended gas source localisation experiment. For other experiments, more sophisticated techniques to determine the robots position might be necessary, or it might be adequate to rely solely on the odometry. Then, the odometry correction chain would have to be replaced by an extended version, or both algorithms that use the corrected position information would have to be connected directly to the odometry object. Otherwise, no further changes would have to be done, because in either case the position information is accessed using the same interface.

A.3   Summary

The software framework introduced in this chapter is designed to support the process of developing robot control applications. The main idea is to map the functional units of an application to objects and to model the cooperation between these objects by dynamically configurable data flow chains. A data flow chain is represented by a cascade of connected objects, which are updated downstream if a certain object changes its internal state. The possibility of adjusting the timing of particular data flow chains is provided by a latency period that can be assigned to each object, meaning that the object cannot trigger an update cascade before this period has elapsed. Thus, a maximum update frequency can be specified for each part of the data flow chains. Due to this feature the framework is called DDFlat (Dynamic Data Flow with Latency). A latency-based mechanism was chosen to avoid a restriction to real-time operating systems (RTOS).

DDFLat enables fast development while keeping the functional parts of the program clearly separated. The basic functionality to maintain the data flow in an application is implemented with regard to an efficient interaction between the involved DDFLat objects. The decoupling of the functional building blocks supports reusability of components and provides a demonstrative way to visualise how the application works. Further on, it facilitates debugging and testing (by means of supporting data logging and simulation) for applications that have to interact with a realistic environment whose dynamics cannot be halted like in conventional debuggers.

Due to the mentioned advantages, the DDFLat framework is of interest for other application domains, too. Consequently, also the programs required for this thesis, which were needed for other purposes than robot control, were implemented using this framework. As an example, the software for the absolute positioning system W-CAPS, is detailed in Appendix B.

Future work should address primarily the implementation of a possibility to place different data flow chains into different processes or concurrent threads of control [NBF96]. With the current version of DDFLat, data flow chains are executed serially without a possibility to interleave execution of different chains. Prioritising a particular task is therefore feasible only in a limited way. More important tasks can be arranged such that they are executed first after a data flow chain was triggered by a device object, for example. However, all of the remaining tasks have to be processed before the high priority task can be executed again. Though the corresponding problems can be avoided in most of the cases by carefully choosing the latency times, a thread-based version would be preferable, which allows to assign DDFLat objects to threads that offer a scheduling priority. In a thread-based version, on the other hand, the problem of how to share data objects properly among different threads would have to be addressed. While data objects are not supposed to be modified by subsequent algorithms, a thread-based framework would have to avoid the situation where a particular data object is modified (from an algorithm connected upstream) before each algorithm downstream is processed that might access it.

Other extensions could be incorporated in order to facilitate the software developing process. This includes the implementation of a tool to support adding of new classes (the steps shown in Fig. A.6) by generating class skeletons in a wizard-like way. Further on, the composition of DDFLat applications could be eased by a graphical user interface to assemble source code similar to Listing A.1. Inversely, it would be also possible to implement a parser to generate an illustration of the process flow automatically from the source code by means of a DDFLat diagram.

Appendix B

The Absolute Positioning System W-CAPS

For experiments in robotics it is frequently necessary (and often helpful) to possess a method that allows investigators to determine the absolute position of a robot or other moving objects.

Available positioning systems most often apply trilateration to determine 2D or 3D coordinates of the tracked object. Frequently time-of-flight measurements (of ultrasonic waves or laser light) are utilised to provide the distance measurements. While ultrasonic systems offer a low cost but not very accurate solution (see for example [RM01]), optical systems that use laser light are comparatively expensive. An overview of ultrasonic and optical positioning systems is given in [Eve95]. An alternative solution that is not subject to the problem of obstructed lines of sight is provided by electromagnetic systems. Here, the magnetic field generated by fixed coils (beacons) is sampled by a mobile sensor unit to determine the distance to the coils [PH00]. Such systems require, however, strong magnetic fields and are susceptible to interference from metallic objects or magnetic fields. Accordingly, sophisticated techniques are required to compensate for distortions and consequently commercially available systems (for example "Liberty" from Polhemus [Pol] or "Flock of Birds" from Ascension [ATC]) are also rather expensive.

This chapter presents an inexpensive vision-based positioning system based on triangulation, which was used for the experiments presented in Chapter 5 and Chapter 6. It provides reliable and accurate measurements by tracking a distinctly coloured object. In order to reduce costs, the system uses a number of web-cameras to acquire images from different angles. The object chosen to be tracked was a coloured "hat" made of cardboard, which can be worn by a person (see Fig. B.1) or placed on top of a robot (see Figs. B.3 and 3.1).

The rest of this chapter is structured as follows. First, the set-up of the positioning system is introduced in Section B.1. Then, a detailed description of the applied method to determine the 2D coordinates of the tracked object is given (Section B.2) and example applications are presented (Section B.4). Finally, conclusions and suggestions for future work are given in Section B.5.

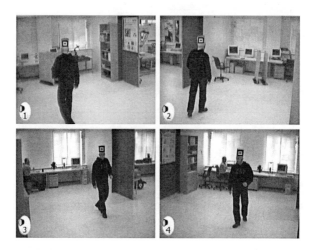

Figure B.1: *Image samples of the four web-cameras used. The computed centre of the coloured "hat" that is tracked by the absolute positioning system, is indicated on top of the raw images.*

B.1   Set-Up

The web-camera-based absolute positioning system ("W-CAPS") was developed such that it can be utilised with an arbitrary number of cameras ($N \geq 2$). It is programmed using the DDFLat framework, which is discussed in Appendix A. A variable number of cameras is supported by the DDFLat framework insofar as the objects, which represent the calculations required for each camera can simply be copied in order to add more cameras. To achieve the results presented here, four Philips PCVC 740K web-cameras were used with a resolution of 320 × 240 pixels. These cameras were mounted at a height of approximately 2 m in the corners of the 10.6 × 4.5 m laboratory room shown in Fig. B.4. The orientation and position was adjusted to cover a large area of interest with as many cameras as possible. All the calculations were performed on a Pentium III PC, which was connected to the web-cameras by a 4 × USB port.

The whole system can be arranged quickly because it uses standard components that should be easily available: first, the webcameras and the (USB) connectors as well as a standard PC with a sufficient number of free (USB) ports is needed. Next, a distinctly coloured object is necessary, which can be assembled using coloured cardboard. Further, a stable support is often needed to mount the cameras. To attach the cameras, the W-CAPS installations arranged so far use either already

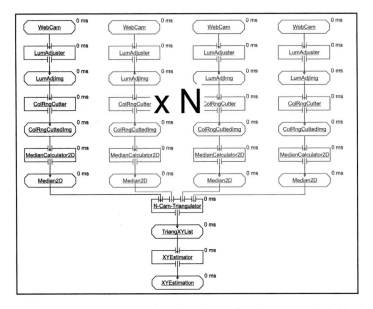

Figure B.2: *DDFLat diagram that shows the mode of operation of W-CAPS.*

available surfaces in the room, or supporting wooden plates, which were screwed to the wall.

B.2   Determining the 2D Coordinates

In order to determine the position of the coloured object, the angle φ_i, at which the centre of the coloured object appears, is first determined for each camera. For every combination of two cameras i, j that both actually sense the whole coloured object, an estimate of the 2D position $\vec{x}_{ij}$ is then calculated by triangulation. Using N cameras up to $N(N-1)/2$ valid position estimates result from each snapshot taken, which are combined to determine the final estimate $\vec{x}$. The individual steps that are executed in order to calculate a position estimate $\vec{x}$ correspond to DDFLat algorithm classes. These steps can be traced in the DDFLat diagram of W-CAPS, which is shown in Fig. B.2. Additionally, a sequence of images visualising the intermediate results is depicted in Fig. B.3. Note that the streaming objects to store the intermediate results are not indicated in Fig. B.2.

To compensate for different lighting conditions, the original colour values (r,g,b) are first normalised by the algorithm `LumAdjuster` as

$$(r', g', b') = \begin{cases} \frac{255}{r+g+b}(r, g, b) & \text{if } r + g + b \geq B_{norm} \\ (r, g, b) & \text{if } r + g + b < B_{norm} \end{cases} . \tag{B.1}$$

Thus, the relative strength of the dominant colour channel is amplified. The threshold B_{norm} is used to prevent amplification of noise in dark regions (examples of normalised images are shown in the second column of Fig. B.3).

Next, pixels within a given contiguous rgb-colour range are selected by the algorithm `ColRngCutter` as

$$(r', g', b') \rightarrow \begin{cases} 1 & \text{if } (r', g', b') \in \Gamma \\ 0 & \text{otherwise} \end{cases} , \tag{B.2}$$

where the color range Γ is given by:

$$\Gamma = [(r_{min}, g_{min}, b_{min}), (r_{max}, g_{max}, b_{max})]. \tag{B.3}$$

Examples of the extracted area are shown in the third column in Fig. B.3.

Then, the median values (X_i, Y_i) of the corresponding pixel-coordinates are calculated for each camera i by the algorithm `MedianCalculator2D`. In order to ensure that the centre of the coloured object is found correctly, the size of the colour blob and whether it lies completely inside the picture is checked. This is done by verifying the following heuristic conditions: first, the number n_{arr} of rows that contain at least N_{row} successive pixels with the tracked colour is evaluated. The median values are calculated if n_{arr} exceeds a certain threshold N_{arr}. This condition is checked for rows only, because just the x-coordinate of the median X_i is used to calculate the angle φ_i at which the centre of the coloured object appears. Second, the border of the image is checked to avoid using a colour blob that is partially outside the image. Because the centre of the visible part of the coloured area would not be valid in such cases, no median values are calculated if there are less than N_{col} empty columns between the median value X_i and the border next to it. A column is considered empty if it contains less than N_{void} pixels of the tracked colour. Again, this condition is applied to the x-axis only. The position of the calculated median is indicated on top of the raw image in the rightmost column of Fig. B.3. As a suitable set of parameters, $N_{row} = 3$, $N_{arr} = 15$, $N_{col} = 3$ and $N_{void} = 3$ were found in initial tests. This parameter set was used in all the experiments reported in this thesis.

In the next step, the angle of the centre of the colour blob φ_i is calculated for each camera i from the pixel position $n_{X,i}$ corresponding to the median value X_i using:

$$\varphi_i = \alpha_i - n_{X,i} \cdot \frac{\Delta\alpha_i}{n_{X,res}}, \tag{B.4}$$

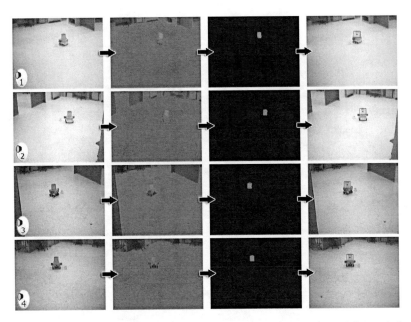

Figure B.3: *Intermediate results during the calculation of the median of the tracked colour blob. The first column displays the raw image of all the 4 cameras used. Then, the raw image with normalised colours (second column) and the area that is assigned to the tracked colour (third column) are shown. Finally, the computed median is indicated on top of the raw image in the rightmost column.*

where $n_{X,res}$ is the horizontal resolution of the web-cameras and $\Delta\alpha_i$, α_i are the covered angle and the angle that corresponds to the left edge in the images of camera i, respectively.

Finally, a list of positions is triangulated for all combinations of two cameras that detected the hat. To avoid ambiguous results, only combinations φ_i, φ_j are considered for which the direction differs sufficiently. Thus, estimates $\vec{x}_{ij}$ are calculated as

$$\vec{x}_{ij} = \frac{(C_iB_j - C_jB_i, \ A_iC_j - A_jC_i)}{A_iB_j - A_jB_i} \qquad (B.5)$$

$$A_i = sin(\varphi_i), \ B_i = -cos(\varphi_i), \ C_i = A_iX_i + B_iY_i, \qquad (B.6)$$

if the difference of the direction

$$\delta_{dir}(\varphi_i, \varphi_j) = min(|\varphi_i - \varphi_j|, |\varphi_i - \varphi_j \pm \pi|) \qquad (B.7)$$

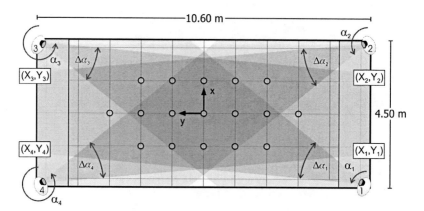

Figure B.4: *Floor plan of a laboratory room where the absolute positioning system W-CAPS was installed. Also shown are the positions that were used for calibration.*

exceeds φ_{min}. These calculations are performed by the DDFLat algorithm N-Cam-Triangulator, which generates a list of the estimates $\vec{x}_{ij}^t$ for the current time step t. This list might have from zero to $N(N-1)/2$ entries.

Ultimately, the overall estimate $\vec{x}_t$ is calculated by averaging over valid estimates $\vec{x}_{ij}^t$ by the algorithm XYEstimator. In order to ensure the validity of the estimates, an outlier detection is carried out first as follows: the position of the colour blob is propagated using the last valid estimate $\vec{x}_{last}$ and the speed $\vec{v}_{last}$, which is determined from the most recent valid positions. An estimate $\vec{x}_{ij}^t$ is believed to be valid if it lies inside a circle with radius $r(t)$ around the propagated position. The radius of the circle is increased linearly by $v_{max} \cdot (t - t_{last})$ to enable recovery in the case of lost positions, using an assumed maximum speed v_{max} and the time since the last valid estimate was detected $(t - t_{last})$.

$$(x,y)_t = \overline{(x,y)_{ij}^t} \text{ with } i,j \in \{1,...,N\}, i \neq j$$
$$\text{and } \vec{x}_{ij}^t \in Circle(\vec{x}_{last} + \vec{v}_{last}, v_{max} \cdot (t - t_{last})) \tag{B.8}$$

B.3   Calibration

The parameters of the cameras (heading α_i, coordinates X_i, Y_i and angular range $\Delta\alpha_i$) are determined by an initial calibration process. This step is crucial because the estimation performance of the whole system depends heavily on the accuracy of the camera parameters.

First, the values of the pixel coordinates $n_{X,i,k}^{(l)}$ of the centre of the hat are determined from K images taken with each camera $i \in 1, \ldots N$ for L known positions $\vec{p}^{(l)}$ of the coloured object. With this set of input data, position estimates $\vec{x}_{i,k}^{(l)}$ are calculated according to equations B.4 – B.8 using a particular set of parameters $\{\alpha_i, X_i, Y_i, \Delta\alpha_i\}$. Finally, the average distance $\bar{d}$ between the estimated and the known positions, calculated as

$$\bar{d} = \sum_{k=1}^{k=K} \sum_{l=1}^{l=L} |\vec{x}_{i,k}^{(l)} - \vec{p}^{(l)}|, \qquad (B.9)$$

is minimised. Any optimisation technique might be used for this purpose. Here, a hillclimbing algorithm was applied starting from a reasonable set of parameters determined by hand. It was found to be advantageous to start optimising just the heading of the cameras α_i first, considering the other parameters as fixed. Then, if no improvement is possible any more, all of the parameters are optimised.

Despite the comparatively poor horizontal resolution of 320 pixels, a good accuracy of $\bar{d} \approx 1$ cm could be achieved using the 17 positions for calibration shown in Fig. B.4.

B.4 Example Applications

The absolute positioning system W-CAPS was used in a number of experiments with the Koala robot and the Mark III Mobile Nose (see Fig. 3.1), which is introduced in Section 3.2. These experiments are discussed in Chapter 5 (gas distribution mapping) and Chapter 6 (reactive gas source localisation), respectively.

Tracking a mobile robot enables improvement of the accuracy of W-CAPS for two reasons: first, the assumed maximum speed v_{max} used in Eq. B.8 is well known and thus outliers can be detected more reliably. Second, odometry information provided by the robot can be fused. In the experiments mentioned, this was done by calculating a new position estimate (using the last estimate and relative odometry information) and adding the low pass filtered deviation of this estimate from the absolute position measured by the camera system. Examples where odometry information was used to determine the position of a comparatively slow robot (v_{max} could be set to 15 cm/s) are shown in Figs. 6.4 – 6.7 and in Appendix E.

Additionally, the information about the heading of the robot provided by its odometry can be improved using the positioning system. While moving with a non-zero translational speed, information about the current heading ϑ can be obtained by calculating the *measured tangent* to the robot's path as

$$\Delta \vec{p}_t^{mt} = \vec{p}_t - \vec{p}_{t-1}. \qquad (B.10)$$

Because of its derivative character the headings calculated with Eq. B.10 provide a rather noisy series of estimates, especially when the robot is moving at low translational speeds. Nevertheless the calculated values can be used to compensate for long term odometry drift. This was done by fusing the heading estimates derived from the measured tangent $\hat{\vartheta}_t^{mt}$ with those from odometry $\hat{\vartheta}_t^{odo}$ using a heuristic nonlinear filter described by the following equations:

$$\hat{\vartheta}_t = \frac{\gamma_t \hat{\vartheta}_t^{odo} + \hat{\vartheta}_t^{mt}}{\gamma_t + 1}, \tag{B.11}$$

$$\hat{\vartheta}_t^{odo} = \hat{\vartheta}_{t-1} + \vartheta_t^{odo} - \vartheta_{t-1}^{odo}, \tag{B.12}$$

$$\gamma_t = \begin{cases} \gamma_{min} & \text{if } \delta\hat{\vartheta}_t < \Delta\vartheta_{min} \\ \gamma_{max} & \text{if } \delta\hat{\vartheta}_t > \Delta\vartheta_{max} \\ \gamma_{min} + (\gamma_{max} - \gamma_{min})\frac{\delta\hat{\vartheta}_t - \Delta\vartheta_{min}}{\Delta\vartheta_{max} - \Delta\vartheta_{min}} & \text{otherwise} \end{cases} \tag{B.13}$$

$$\delta\hat{\vartheta}_t = |\delta_{ang}(\hat{\vartheta}_t^{Odo}, \hat{\vartheta}_t^{MT})|, \tag{B.14}$$

where the function $\delta_{ang}(\vartheta_i, \vartheta_j)$ returns the angular difference between two angles ϑ_i and ϑ_j.

This means that headings of the measured tangents that differ greatly from the odometry estimate, which is calculated by adding the last estimate and differential odometry information, are considered to be unreliable and are thus integrated with a smaller weight. During the experiments the following parameters were used, which were found to yield good results in initial tests: $\Delta\vartheta_{min} = 5.0°$, $\Delta\vartheta_{max} = 180.0°$, $\gamma_{min} = 3.0$, $\gamma_{max} = 100.0$.

B.5   Conclusions and Future Work

The positioning system introduced in this chapter was developed because it was required for some of the experiments in this work and not primarily as a self-contained research issue. Rather than aiming at an optimal solution, a straightforward one that satisfies the given requirements on accuracy and robustness was therefore chosen. Consequently, there are several issues remaining to be done and also possible extensions that can improve the performance of the system, which are not implemented yet. First of all, a proper evaluation of the accuracy of the position information is clearly needed. Because an alternative measuring system to determine the "ground truth" is not available, the error has to be estimated from the position information provided by W-CAPS only. This can be done, for example, by smoothing the data obtained from a slowly moving platform with a large Gaussian kernel and then comparing single samples with the smoothed version of the signal [BV03]. Such an investigation would have to address especially the dependency between the accuracy obtained and the number of cameras used.

Several extensions of W-CAPS are possible that are likely to improve the performance of the system. In order to increase the robustness concerning different lighting conditions, for example, it would be probably advantageous to define the tracked colour as a region in an alternative, perception-oriented colorspace like HSV [Smi78] or HSI [GW92]. In this regard, it might be also advantageous to study other than cuboid-shaped regions. With the applied normalisation (Eq. B.1), however, a fairly lighting-independent behaviour was achieved. The positioning system could be utilised with the chosen colour (bright green) in the night as well as during daytime using a single hand-crafted rgb-region (Eq. B.3), whereas the blinds had to be closed in case of direct sunlight irradiation.

As concerning the robustness, several extensions of W-CAPS are also imaginable to improve the accuracy. This includes a more sophisticated way of infering the final estimate from the available knowledge (prior estimates and odometry information) by using a Kalman filter [May79]. Moreover, it is probably beneficial to introduce a certainty measure for single position estimates. Up to now, W-CAPS calculates an estimate by averaging over those triangulated positions that are considered reliable. Future work might therefore investigate whether the accuracy of the system can be increased by assigning different weights to individual triangulation results. These weights might be related either to the total distance to the cameras used or to the angular difference between them.

Another issue that could be addressed is the run time of W-CAPS. The DDFLat framework was used in order to enable fast development while providing an efficient data flow process. Nevertheless, there is a considerable potential to optimise the efficiency of the DDFLat objects involved, for example by combining the selection of the tracked colour and the calculation of the median in a single class. Rather than the computing power, however, the limiting factor in the currently used set-up, is the bandwidth of the USB 1.0 port resulting in an update frequency of approximately 1.25 Hz.

Finally, the minimisation algorithm used for calibration could be improved in order to achieve a more user-friendly handling. With the applied hillclimbing algorithm (see Section B.3), it depends strongly on the quality of the initial guess of the parameters whether a reasonable parameter set is found. While a sufficiently accurate initial guess can be determined by hand and by applying the optimisation step iteratively regarding an increasing number of calibration points, it would be preferable to implement a global search technique, like a genetic algorithm [Hol75; Gol98], for example.

Though the implementation cannot be considered as technically mature at the moment, W-CAPS has been utilised successfully in several experiments. This includes the experiments on gas distribution mapping (Chapter 5) and reactive gas source localisation (Chapter 6) presented in this thesis. Further on, W-CAPS has been used to track the position of people (wearing the cardboard hat) in order to train a neural net for person tracking [CMH+03]. In addition, it is currently being adapted by the RoboCup team TeamSweden [SBJW02], in order to provide ground truth information about the position of their legged robots.

Appendix C

Properties of the Concentration Gridmap Weighting Function

C.1 Variation Along the Driven Path

In this section, the weighting function (Eq. 5.5), which is calculated in the process of creating a concentration gridmap, is analysed with respect to its course along the driven path. Therefore, a trajectory is considered that contains an infinite series of equidistant point measurements:

$$\vec{x}_i = (i \cdot \Delta x, 0) \qquad i \in \{-\infty, ..., -1, 0, 1, ..., \infty\}. \tag{C.1}$$

Expressing the step width Δx as a fraction η of the width of the Gaussian weighting function σ (see Eq. 5.1), Equation C.1 becomes

$$\vec{x}_i = (i \cdot \eta\sigma, 0) \qquad i \in \{-\infty, ..., -1, 0, 1, ..., \infty\}, \tag{C.2}$$

$$\Delta x = \eta\sigma. \tag{C.3}$$

Hence, the total sum W is given by

$$W(x, y; \sigma) = \frac{1}{2\pi\sigma^2} \sum_{i=-\infty}^{\infty} e^{-\frac{(x - i \cdot \eta\sigma)^2 + y^2}{2\sigma^2}}. \tag{C.4}$$

The minimum and maximum of the projection at $y = 0$ is located for symmetry reasons at $\vec{x}_0 = (0, 0)$ and $\vec{x}_{1/2} = (\frac{1}{2}\eta\sigma, 0)$. Evaluating Eq. C.4 for these locations, yields

$$W(x_0, 0; \sigma) = \frac{1}{2\pi\sigma^2} \Big\{ \sum_{i=-\infty}^{\infty} e^{-\frac{1}{2}\eta^2 i^2} \Big\} = \frac{1}{2\pi\sigma^2} \Big\{ 1 + 2\sum_{i=1}^{\infty} e^{-\frac{1}{2}\eta^2 i^2} \Big\}, \tag{C.5}$$

and

$$W(x_{1/2}, 0; \sigma) = \frac{1}{2\pi\sigma^2} \Big\{ \sum_{i=-\infty}^{\infty} e^{-\frac{(\frac{1}{2}\eta\sigma - i\eta\sigma)^2}{2\sigma^2}} \Big\}$$

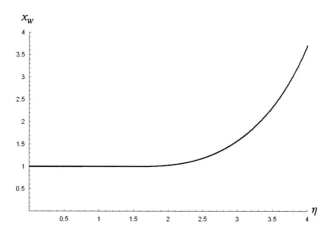

Figure C.1: *Variation of the total weight along the driven path.*

$$= \frac{1}{2\pi\sigma^2}\Big\{ \sum_{i=-\infty}^{\infty} e^{-\frac{1}{8}\eta^2} e^{-\frac{\frac{1}{2}\eta^2(i^2-i)}{2\sigma^2}} \Big\}$$

$$= \frac{e^{-\frac{1}{8}\eta^2}}{2\pi\sigma^2}\Big\{ 1 + \sum_{i=1}^{\infty} e^{-\frac{1}{2}\eta^2(i^2-i)} + \sum_{i=1}^{\infty} e^{-\frac{1}{2}\eta^2(i^2+i)} \Big\}$$

$$= \frac{e^{-\frac{1}{8}\eta^2}}{2\pi\sigma^2}\Big\{ 1 + 2\sum_{i=1}^{\infty} e^{-\frac{1}{2}\eta^2 i^2} cosh(\frac{1}{2}\eta^2 i) \Big\}. \tag{C.6}$$

The infinite series in Equations C.5 and C.6 can be expressed using the theta function $\theta_a(u, q)$ [BSMM99]:

$$W(x_0, 0; \sigma) = \frac{1}{2\pi\sigma^2}\vartheta_3(0, e^{-\frac{1}{2}\eta^2}), \tag{C.7}$$

$$W(x_{1/2}, 0; \sigma) = \frac{1}{2\pi\sigma^2}\vartheta_2(0, e^{-\frac{1}{2}\eta^2}), \tag{C.8}$$

The theta function converges for $q < 1$. Consequently, the quotient of Equations C.7 and C.8

$$x_W = \frac{W(x_0, 0; \sigma)}{W(x_{1/2}, 0; \sigma)} = \frac{\vartheta_3(0, e^{-\frac{1}{2}\eta^2})}{\vartheta_2(0, e^{-\frac{1}{2}\eta^2})}. \tag{C.9}$$

exists for $\eta > 0$. This function is shown in Fig. C.1. The value x_W increases monotonically with η, indicating the tendency that the Gaussians do not sum up to a constant value in-between their centres if the distance between these centres becomes too big. However, the rise in x_w is comparatively weak for $\eta < C_{x_W}$ with a constant between approximately $C_{x_W} = 1.5$ ($x_W(1.5) = 1.00062$) and $C_{x_W} = 2.5$ ($x_W(2.5) = 1.18579$):

$$x_W \approx 1 \qquad \text{for} \qquad 0 < \eta < C_{x_W}. \tag{C.10}$$

Using the definition of η in Eq. C.3, a relation between the width of the Gaussian density function σ and the step width can be devised as

$$x_W \approx 1 \qquad \text{for} \qquad 0 < \Delta x < C_{x_W}\sigma. \tag{C.11}$$

C.2 Width Orthogonal to the Driven Path

In this section, the weighting function W_t (Eq. 5.5) used to create concentration gridmaps is analysed with respect to its course perpendicular to the driven path. Again, a trajectory is considered that contains an infinite series of equidistant point measurements as defined in the equations C.2 and C.3. For locations at the trajectory ($y = 0$), W_t is given as

$$W(x, y = 0; \sigma) = \frac{1}{2\pi\sigma^2} \sum_{i=-\infty}^{\infty} e^{-\frac{(x-i\cdot\eta\sigma)^2}{2\sigma^2}}, \tag{C.12}$$

while the value for corresponding locations perpendicular to the path is

$$W(x, y = \Delta y; \sigma) = \frac{1}{2\pi\sigma^2} \sum_{i=-\infty}^{\infty} e^{-\frac{(x-i\cdot\eta\sigma)^2+y^2}{2\sigma^2}} = \frac{e^{-\frac{\Delta y^2}{2\sigma^2}}}{2\pi\sigma^2} \sum_{i=-\infty}^{\infty} e^{-\frac{(x-i\cdot\eta\sigma)^2}{2\sigma^2}}. \tag{C.13}$$

Consequently, the decay perpendicular to the path as

$$\frac{W(x, y = \Delta y; \sigma)}{W(x, y = 0; \sigma)} = e^{-\frac{\Delta y^2}{2\sigma^2}} \tag{C.14}$$

does not depend on η, and thus is also independent of the step witdh Δx.

Appendix D

Concentration Mapping Experiments

This section presents selected snapshots of concentration gridmaps of all the mapping trials performed. Concentration mapping is introduced and discussed in Chapter 5. Within this chapter also the applied data acquisition strategies are explained (see Sections 5.7.1 – 5.7.3).

To indicate concentration values, all the concentration gridmaps shown in this work are displayed using different shadings of grey whereas dark shadings correspond to low and light shadings to high relative concentrations. The values higher than 90% of the maximum are indicated with a second range of dark-to-light shadings {of red}. Unexplored cells are displayed with another colour {green}. The chosen illustration utilises the full range of available shadings in order to visualise relative intensities in a particular map. Consequently, the absolute value of a cell in two different gridmaps cannot be compared directly by means of comparing the corresponding brightness in the illustration. Further, the maximal and the 90% median cell are marked in the concentration gridmaps. A {blue} dot refers to the the maximal cell (the cell with the highest concentration value), which is located in the brightest part of the second range of dark-to-light shadings {of red}. Finally, the 90% median (the median of the x- and y- coordinates of the area, which is defined by those cells with a value of at least 90% of the maximum) is indicated by another {bright green} dot.

D.1 Predefined Path – Rectangular Spiral

RectSpiral-1: source position: (0.0 m, 0.0 m), duration: 175 min

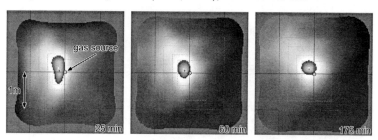

RectSpiral-2: source position: (0.0 m, 0.0 m), duration: 177 min

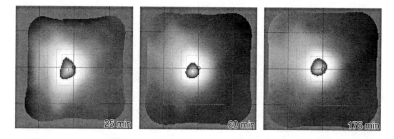

RectSpiral-3: source position: (0.0 m, 0.0 m), duration: 184 min

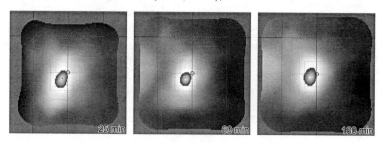

Figure D.1: *Snapshots of the concentration gridmap obtained in experiments "RectSpiral-1" – "RectSpiral-3"*

RectSpiral-4: source position: (0.0 m, 0.0 m), duration: 189 min

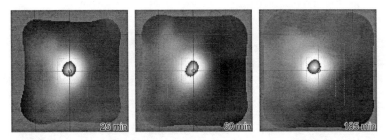

Figure D.2: *Snapshots of the concentration gridmap obtained in the experiment "RectSpiral-4"*

D.2 Predefined Path – Sweeping Movement

Sweeping-1: source position: (0.0 m, -0.80 m), duration: 160 min

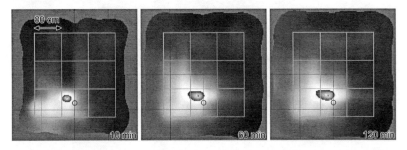

Sweeping-2: source position: (-0.80 m, -0.80 m), duration: 151 min

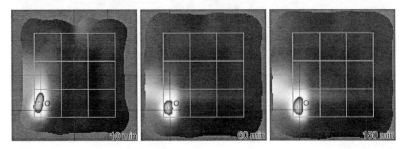

Figure D.3: *Snapshots of the concentration gridmap obtained in experiments "Sweeping-1" – "Sweeping-2"*

Sweeping-3: source position: (0.80 m, -0.80 m), duration: 152 min

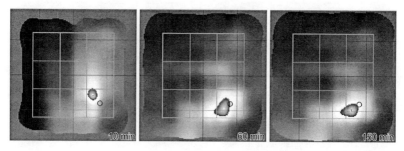

Sweeping-4: source position: (0.80 m, 0.0 m), duration: 151 min

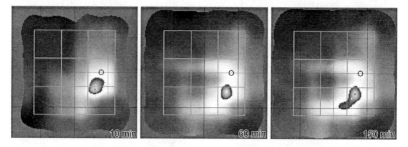

Sweeping-5: source position: (-0.80 m, 0.0 m), duration: 147 min

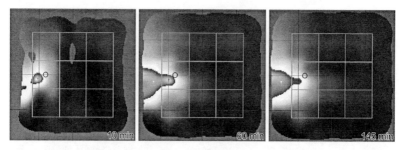

Figure D.4: *Snapshots of the concentration gridmap obtained in experiments "Sweeping-3" – "Sweeping-5"*

Sweeping-6: source position: (0.80 m, 0.0 m), duration: 151 min

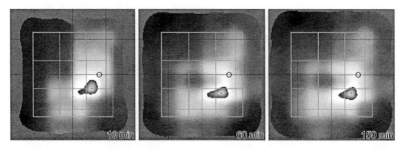

Sweeping-7: source position: (0.80 m, 0.80 m), duration: 152 min

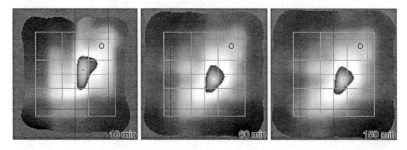

Sweeping-8: source position: (-0.80 m, 0.80 m), duration: 152 min

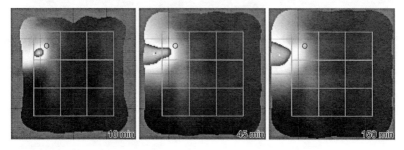

Figure D.5: *Snapshots of the concentration gridmap obtained in experiments "Sweeping-6" – "Sweeping-8"*

D.3   Reactive Gas Source Tracing

D.3.1   Experiments with Uncrossed Inhibitory Connections ("Permanent Love")

PL-Mid1: source position: (0.0 m, 0.0 m), duration: 241 min, $K_v = 5$ cm/s

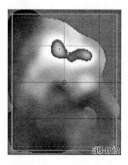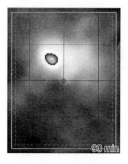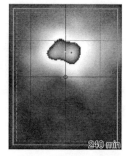

PL-Mid2: source position: (0.0 m, 0.0 m), duration: 187 min, $K_v = 5$ cm/s

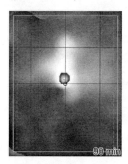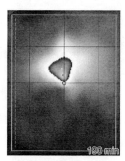

Figure D.6: *Snapshots of the concentration gridmap obtained in experiments "PL-Mid1" – "PL-Mid2"*

PL-Mid3: source position: (0.0 m, 0.0 m), duration: 175 min, $K_v = 5$ cm/s

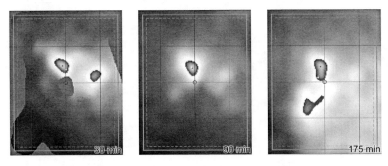

PL-UL1: source position: (-1.15 m, +1.50 m), duration: 191 min, $K_v = 5$ cm/s

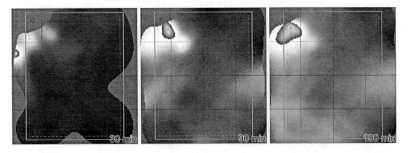

PL-UR1: source position: (+1.15 m, +1.50 m), duration: 187 min, $K_v = 5$ cm/s

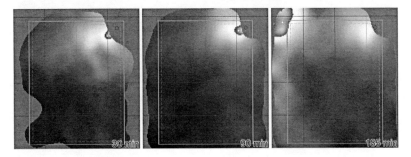

Figure D.7: *Snapshots of the concentration gridmap obtained in experiments "PL-Mid3", "PL-UL1" and "PL-UR1"*

PL-LR1: source position: (+1.15 m, -1.50 m), duration: 190 min, $K_v = 3$ cm/s

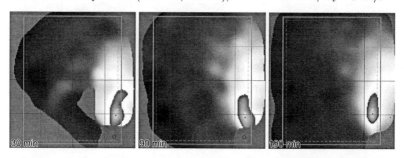

PL-LR2: source position: (+1.15 m, -1.50 m), duration: 187 min, $K_v = 3$ cm/s

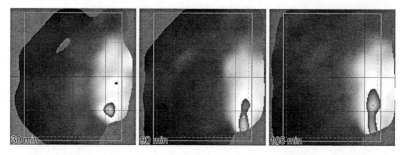

PL-LL1: source position: (+1.15 m, -1.50 m), duration: 133 min, $K_v = 3$ cm/s

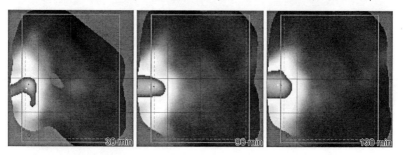

Figure D.8: *Snapshots of the concentration gridmap obtained in experiments "PL-LR1", "PL-LR2" and "PL-LL1"*

D.3.2 Experiments with Crossed Inhibitory Connections ("Exploring Love")

EL-Mid1: source position: (0.0 m, 0.0 m), duration: 189 min, $K_v = 5$ cm/s

EL-Mid2: source position: (0.0 m, 0.0 m), duration: 181 min, $K_v = 3$ cm/s

  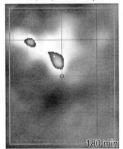

EL-Mid3: source position: (0.0 m, 0.0 m), duration: 186 min, $K_v = 3$ cm/s

Figure D.9: *Snapshots of conc. gridmaps, experiments "EL-Mid1" – "EL-Mid3"*

Appendix E

Braitenberg-Type Experiments

This section presents the path plots of all the gas source tracing experiments with an active source where the robot was steered as a Braitenberg vehicle. Details of the applied strategy are given in Section 6.2. The experimental set-up is explained in Section 6.4 and the testing process in Section 6.4.1. Further on, the results are discussed in Section 6.5.1 and Section 6.5.2, respectively.

Each path plot shows the position of the robot's centre and its front corners indicated by small dots. Furthermore, the virtual repellent walls are indicated by two broken lines that enclose the area where the repellent force increases with the penetration depth of the robot. The clearance area around the gas source, which was either a real source or just an assumed one for reference tests, is indicated by two circles, corresponding to the obstacle radius of the gas source concerning the corners of the robot (inner circle) and its center (outer circle), respectively.

E.1 "Permanent Love" Trials

PL-Mid1: uncrossed inh. connections (1–x), source in the middle

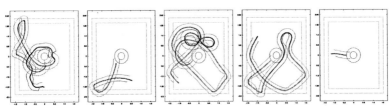

Figure E.1: *Trials 1–5 of experiment PL-Mid1: source position: (0.0 m, 0.0 m), $K_v = 5$ cm/s, duration = 241 min (198 min of tracing), total path length = 383.5 m, source hits = 45, average distance to source = 117.4 cm ± 46.6 cm.*

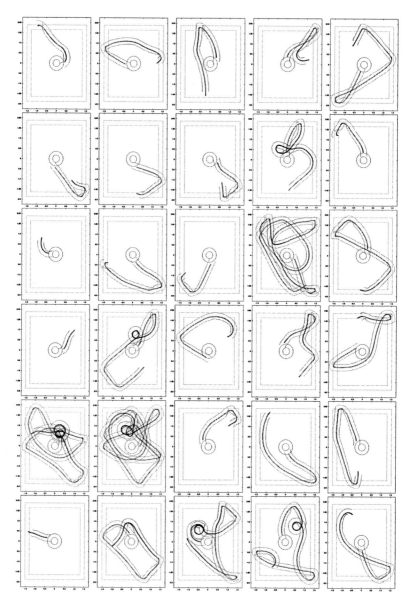

Figure E.2: *Trials 6–35 of experiment PL-Mid1.*

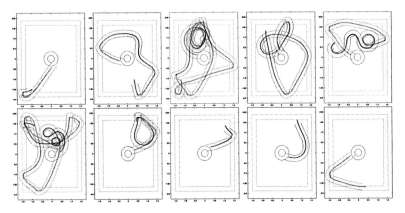

Figure E.3: *Trials 36–45 of experiment PL-Mid1.*

PL-Mid2: uncrossed inh. connections (1–x), source in the middle

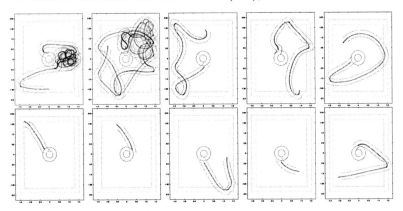

Figure E.4: *Trials 1–10 of experiment PL-Mid2: source position: (0.0 m, 0.0 m),
K_v = 5 cm/s, duration = 187 min (155 min of tracing), total path length = 333.7 m
source hits = 38, average distance to source = 117.5 cm ± 45.7 cm.*

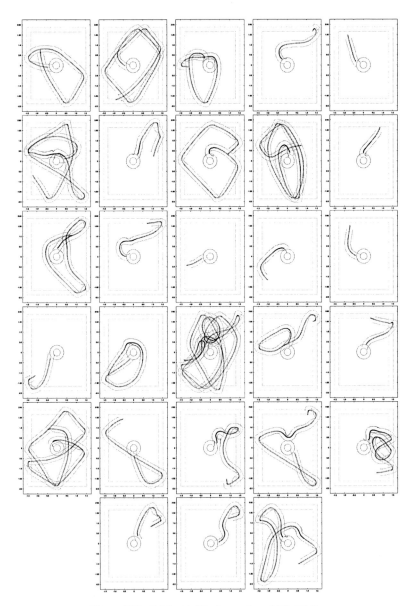

Figure E.5: *Trials 11–38 of experiment PL-Mid2.*

PL-Mid3: uncrossed inh. connections (1–x), source in the middle

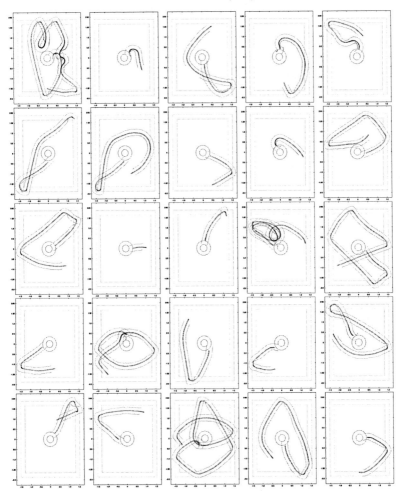

Figure E.6: *Trials 1–25 of experiment PL-Mid3: source position: (0.0 m, 0.0 m), $K_v = 5$ cm/s, duration = 175 min (143 min of tracing), total path length = 326.8 m, source hits = 39, average distance to source = 125.1 cm ± 44.3 cm.*

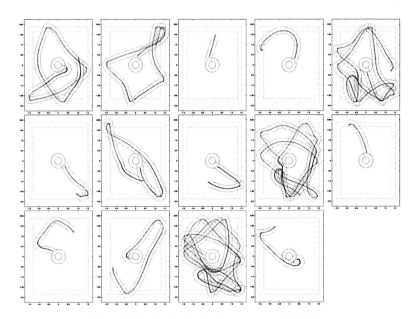

Figure E.7: *Trials 26–39 of experiment PL-Mid2.*

PL-UL1: uncrossed inh. connections (1–x), source in the upper left

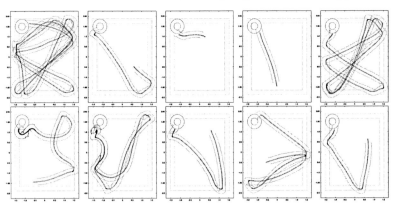

Figure E.8: *Trials 1–10 of experiment PL-UL1: source position: (–1.15 m, +1.5 m), $K_v = 5$ cm/s, duration = 191 min (146 min of tracing), total path length = 344.3 m, source hits = 28, average distance to source = 214.9 cm ± 90.6 cm.*

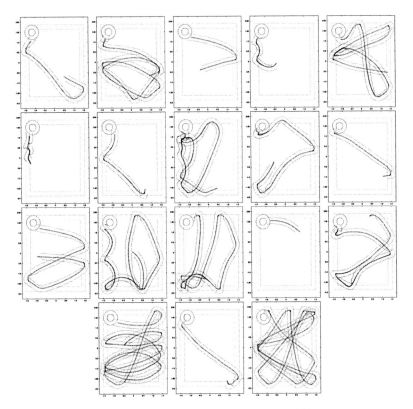

Figure E.9: *Trials 11–28 of experiment PL-UL1.*

PL-UR1: uncrossed inh. connections (1–x), source in the upper right

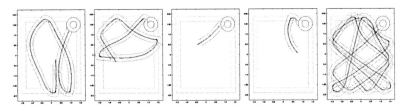

Figure E.10: *Trials 1–5 of experiment PL-UR1: source position: (+1.15 m, +1.5 m), $K_v = 5$ cm/s, duration = 187 min (155 min of tracing), total path length = 389.0 m, source hits = 24, average distance to source = 220.4 cm ± 89.5 cm.*

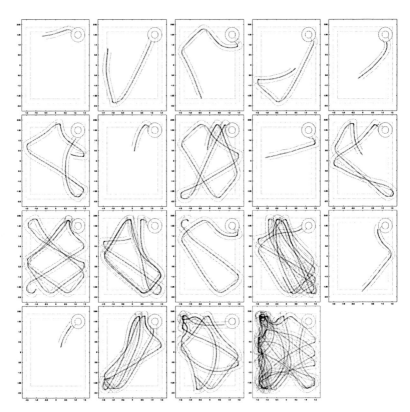

Figure E.11: *Trials 6–24 of experiment PL-UR1.*

PL-LR1: uncrossed inh. connections (1–x), source in the lower right

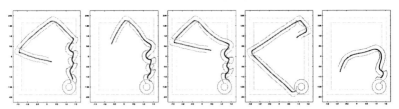

Figure E.12: *Trials 1–5 of experiment PL-LR1: source position: (+1.15 m, -1.5 m),*
$K_v = 3$ *cm/s, duration = 187 min (141 min of tracing), total path length = 188.4 m,*
source hits = 25, average distance to source = 196.0 cm ± 90.8 cm.

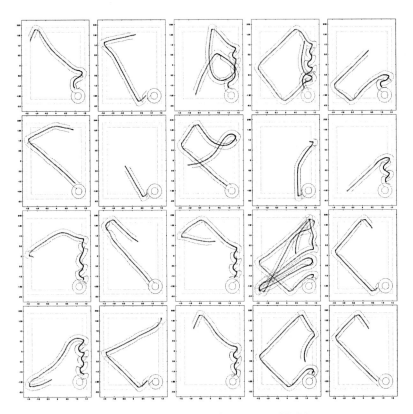

Figure E.13: *Trials 6–25 of experiment PL-LR1.*

PL-LR2: uncrossed inh. connections (1–x), source in the lower right

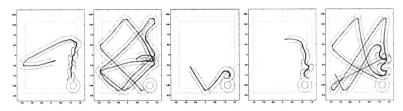

Figure E.14: *Trials 1–5 of experiment PL-LR2: source position: (+1.15 m, -1.5 m),* $K_v = 3$ *cm/s, duration = 187 min (141 min of tracing), total path length = 188.4 m, source hits = 25, average distance to source = 196.0 cm ± 90.8 cm.*

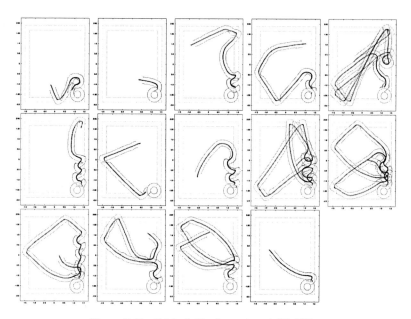

Figure E.15: *Trials 6–19 of experiment PL-LR2.*

PL-LL1: uncrossed inh. connections (1–x), source in the lower left

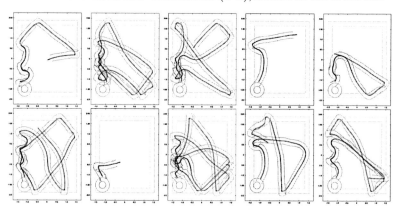

Figure E.16: *Trials 1–10 of experiment PL-LL1: source position: (-1.15 m, -1.5 m), $K_v = 3$ cm/s, duration = 133 min (105 min of tracing), total path length = 137.0 m, source hits = 10, average distance to source = 184.2 cm ± 76.9 cm.*

E.2 "Exploring Love" Trials

EL-Mid1: crossed inh. connections (1–x), source in the middle

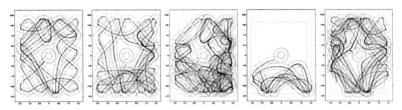

Figure E.17: *Trials 1–5 of experiment EL-Mid1: source position: (0.0 m, 0.0 m), K_v = 5 cm/s, duration = 172 min (169 min of tracing), total path length = 386.3 m, source hits = 5, average distance to source = 137.3 cm ± 40.4 cm.*

EL-Mid2: crossed inh. connections (1–x), source in the middle

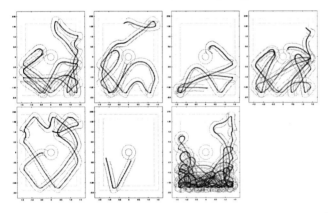

Figure E.18: *Trials 1–7 of experiment EL-Mid2: source position: (0.0 m, 0.0 m), K_v = 3 cm/s, duration = 181 min (175 min of tracing), total path length = 245.9 m, source hits = 7, average distance to source = 143.6 cm ± 38.9 cm.*

EL-Mid3: crossed inh. connections (1–x), source in the middle

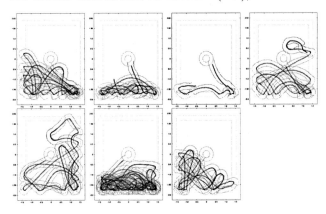

Figure E.19: *Trials 1–7 of experiment EL-Mid3: source position: (0.0 m, 0.0 m),*
$K_v = 3$ *cm/s, duration = 186 min (180 min of tracing), total path length = 230.8 m,*
source hits = 7, average distance to source = 150.3 cm ± 42.5 cm.

Appendix F

Publications From This Thesis

The following publications arose from the work done for this thesis. Most of the published papers are available on-line at

http://www.lilienthals.de/achim/research/

Journal Articles (Main Author)

1. Achim Lilienthal and Tom Duckett, *Building Gas Concentration Gridmaps with a Mobile Robot*, Robotics and Autonomous Systems [LD04a]

2. Achim Lilienthal and Tom Duckett, *Experimental Analysis of Gas-Sensitive Braitenberg Vehicles*, Advanced Robotics [LD04b]

Conference Papers (Main Author)

1. Achim Lilienthal, Holger Ulmer, Holger Fröhlich, Felix Werner, and Andreas Zell *Learning to Detect Proximity to a Gas Source with a Mobile Robot*, IROS 2004 [LUF⁺04b]

2. Achim Lilienthal, Holger Ulmer, Holger Fröhlich, Andreas Stützle, Felix Werner and Andreas Zell, *Gas Source Declaration with a Mobile Robot*, ICRA 2004 [LUF⁺04a]

3. Achim Lilienthal, Denis Reiman and Andreas Zell, *Gas Source Tracing With a Mobile Robot Using an Adapted Moth Strategy*, AMS 2003 [LRZ03]

4. Achim Lilienthal and Tom Duckett, *Approaches to Gas Source Tracing and Declaration by Pure Chemo-Tropotaxis*, AMS 2003 [LD03c]

5. Achim Lilienthal and Tom Duckett, *Creating Gas Concentration Gridmaps with a Mobile Robot*, IROS 2003 [LD03d]

6. Achim Lilienthal and Tom Duckett, *Gas Source Localisation by Constructing Concentration Gridmaps with a Mobile Robot*, ECMR 2003 [LD03f]

7. Achim Lilienthal and Tom Duckett, *Experimental Analysis of Smelling Braitenberg Vehicles*, ICAR 2003 [LD03e] (**Best Paper Award**)

8. Achim Lilienthal and Tom Duckett, *A Stereo Electronic Nose for a Mobile Inspection Robot*, ROSE 2003 [LD03a]

9. Achim Lilienthal and Tom Duckett, *An Absolute Positioning System for 100 Euro*, ROSE 2003 [LD03b]

10. Achim Lilienthal, Andreas Zell, Michael Wandel and Udo Weimar, *Detektion und Lokalisierung einer Geruchsquelle mit einem Autonomen Mobilen Roboter*, Robotik 2002 [LZWW02]

11. Achim Lilienthal, Andreas Zell, Michael Wandel and Udo Weimar, *Experiences Using Gas Sensors on an Autonomous Mobile Robot*, EUROBOT 2001 [LZWW01a]

12. Achim Lilienthal, Andreas Zell, Michael Wandel and Udo Weimar, *Sensing Odour Sources in Indoor Environments Without a Constant Airflow by a Mobile Robot*, ICRA 2001 [LZWW01b]

13. Achim Lilienthal, Andreas Zell, Michael Wandel and Udo Weimar, *Ein Autonomer Mobiler Roboter mit Elektronischer Nase*, AMS 2000 [LZWW00]

Conference Papers (Co-Author)

1. Grzegorz Cielniak, Mihajlo Miladinovic, Daniel Hammarin, Linus Göransson, Achim Lilienthal and Tom Duckett, *Appearance-based Tracking of Persons with an Omnidirectional Vision Sensor*, Omnivis 2003 [CMH+03]

2. Michael Wandel, Achim Lilienthal, Tom Duckett, Udo Weimar and Andreas Zell, *Gas Distribution in Unventilated Indoor Environments Inspected by a Mobile Robot*, ICAR 2003 [WLD+03]

3. Michael Wandel, Achim Lilienthal, Andreas Zell and Udo Weimar, *Mobile Robot Using Different Senses*, ISOEN 2002 [WLZW02]

4. Michael Wandel, Udo Weimar, Achim Lilienthal and Andreas Zell, *Leakage Localisation With a Mobile Robot Carrying Chemical Sensors*, ICECS 2001 [WWLZ01]

Bibliography

[ADDK99] Y. S. Avrithis, A. D. Doulamis, N. D. Doulamis, and S. D. Kollias. A Stochastic Framework for Optimal Key Frame Extraction from MPEG Video Databases. *Computer Vision and Image Understanding*, 75(1/2):3–24, July/August 1999.

[App] AppliedSensor, Linköping (Sweden) formerly MoTech, Reutlingen (Germany). http://www.appliedsensor.com (February 2004).

[ARt] ARtem, Ulm (Germany). http://www.artem.de (February 2004).

[ATC] Vermont Ascension Technology Corporation, Burlington. Flock of Birds Product Brochure. http://www.ascension-tech.com/products/flockofbirds.pdf (February 2004).

[Ate96] J. Atema. Eddy Chemotaxis and Odor Landscapes: Exploration of Nature With Animal Sensors. *Biological Bulletin*, 191(1):129–138, 1996.

[BB] Stadtarchiv Baden-Baden. Aktenzeichen A10/1974, A10/1975, A27/7-625, A27/7-635, A27/7-732, A27/7-734, A27/7-1144, A27/7-1146.

[BBSH59] A. Butenandt, R. Beckmann, D. Stamm, and E. Hecker. Über den Sexuallockstoff des Seidenspinners *Bombyx mori*. Reindarstellung und Konstitution. *Zeitschrift für Naturforschung*, 14b:283–284, 1959.

[BG02] B. S. Blackmore and H. W. Griepentrog. A Future View of Precision Farming. In *Proceedings of PreAgro Precision Agriculture Conference*, pages 131–145, Müncheberg, Germany, Center for Agricultural Landscape and Land Use Research (ZALF), 2002.

[Bis95] Christopher M. Bishop. *Neural Networks for Pattern Recognition*. Oxford University Press Inc., New York, 1995.

[BK88] J. Borenstein and Y. Koren. Obstacle Avoidance with Ultrasonic Sensors. *IEEE Journal of Robotics and Automation*, 4(2):213–218, 1988.

[Bra84] Valentino Braitenberg. *Vehicles: Experiments in Synthetic Psychology*. MIT Press/Bradford Books, 1984.

[Bra85]   J. M. Brady. Artificial Intelligence and Robotics. *Artificial Intelligence and Robotics*, 26:79–121, 1985.

[Bro92]   Rodney Brooks. Artificial Life and Real Robots. In *Proceedings of the First European Conference on Artificial Life (ECAL 1991)*, pages 3–10, Paris, France, 1992.

[BS02]   E. Balkovsky and B. I. Shraiman. Olfactory Search at High Reynolds Number. *Proceedings of the National Acadademy of Sciences of the United States of America (PNAS)*, 99(20):12589–12593, October 2002.

[BS03]   Peter Biber and Wolfgang Straßer. The Normal Distributions Transform: A New Approach to Laser Scan Matching. In *Proceedings of the IEEE/RSJ International Conference on Intelligent Robots and Systems (IROS 2003)*, pages 2743–2748, 2003.

[BSMM99] I. N. Bronstein, K. A. Semendjaev, G. Musiol, and H. Mühlig. *Taschenbuch der Mathematik*. Harri Deutsch, Frankfurt am Main, Thun, 4th edition, 1999.

[BT95]   N. Barsan and A. Tomescu. Calibration Procedure for SnO2-based Gas Sensors. *Thin Solid Films*, 259:91–95, 1995.

[BV03]   J. Bruce and M. Veloso. Fast and Accurate Vision-Based Pattern Detection and Identification. In *Proceedings of the IEEE International Conference on Robotics and Automation (ICRA 2003)*, pages 1277–1282, 2003.

[CAGC94] T.R. Consi, J. Atema, C. Gouldey, and C. Chrysostomidis. AUV Guidance with Chemical Signals. In *Proceedings of the IEEE Symposium on AUV Technology*, pages 450–455, Cambridge, MA, July 1994.

[CK90]   W. P. Carey and B. R. Kowalski. Sensor Arrays for Chemical Analysis of Vapors. *Sensors and Actuators B*, (1):43–47, 1990.

[CM02]   J. Casper and R. Murphy. Workflow Study on Human-Robot Interaction in USAR. In *Proceedings of the IEEE International Conference on Robotics and Automation (ICRA 2002)*, pages 1997–2003, 2002.

[CMH+03] Grzegorz Cielniak, Mihajlo Miladinovic, Daniel Hammarin, Linus Göransson, Achim Lilienthal, and Tom Duckett. Appearance-based tracking of persons with an omnidirectional vision sensor. In *Proceedings of the Fourth IEEE Workshop on Omnidirectional Vision (Omnivis 2003)*, Madison, Wisconsin, 2003.

[CV95]   C. Cortes and V. Vapnik. Support Vector Networks. *Machine Learning*, 20:273–297, 1995.

[DAS01] Tom Duckett, Mikael Axelsson, and Alessandro Saffiotti. Learning to Locate an Odour Source with a Mobile Robot. In *Proceedings of the IEEE International Conference on Robotics and Automation (ICRA 2001)*, pages 4017–4022, 2001.

[Dir] Directed Perception, Burlingame (USA). http://www.dperception.com (February 2004).

[DSCG02] T. Dekker, B. Steib, R. T. Cardé, and M. Geier. L-Lactic Acid: A Human-Signifying Host Cue for the Anthropophilic Mosquito *Anopheles Gambiae. Medical and Veterinary Entomology*, 16:91–98, 2002.

[DTRMS94] R. Deveza, D. Thiel, R. A. Russell, and A. Mackay-Sim. Odor Sensing for Robot Guidance. *The International Journal of Robotics Research*, 3(13):232–239, June 1994.

[Duc00] Tom Duckett. *Concurrent Map Building and Self-Localisation for Mobile Robot Navigation*. PhD thesis, Department of Computer Science, University of Manchester, 2000.

[Duc03] Tom Duckett. A Genetic Algorithm for Simultaneous Localization and Mapping. In *Proceedings of the IEEE International Conference on Robotics and Automation (ICRA 2003)*, pages 434–439, 2003.

[Elf87] A. Elfes. Sonar-Based Real World Mapping and Navigation. *IEEE Transactions on Robotics and Automation*, 3(3):249–265, 1987.

[EM85] A. Elfes and H. P. Moravec. High Resolution Maps from Wide Angle Sonar. In *Proceedings of the IEEE International Conference on Robotics and Automation (ICRA 1985)*, pages 116–121, 1985.

[Eng89] J. F. Engelberger. *Robotics in Service*. Kogan Page, London, 1989.

[Eve95] H. R. Everett. *Sensors For Mobile Robots*. A K Peters Ltd., 6th edition, 1995.

[FD02] Ahmed Mohamod Farah and Tom Duckett. Reactive Localisation of an Odour Source by a Learning Mobile Robot. In *Proceedings of the Second Swedish Workshop on Autonomous Robotics*, pages 29–38, Stockholm, Sweden, October 2002.

[FG40] G.S. Fraenkel and D.L. Gunn. *The Orientation of Animals*. Clarendon Press, Oxford, 1940.

[Fig] Figaro, Osaka (Japan). http://www.figarosensor.com (February 2004).

[FM00] M. O. Franz and H. A. Mallot. Biomimetic Robot Navigation. *Robotics and Autonomous Systems*, 30:133–153, 2000.

[Gag93] Douglas Gage. Randomized Search Strategies with Imperfect Sensors. In *Proceedings of SPIE Mobile Robots VIII*, volume 2058, pages 270–279, Boston, USA, 1993.

[Gar87] Julian W. Gardner. Pattern Recognition in the Warwick Electronic Nose. In *8th Int. Congress of the European Chemoreception Research Organisation*, 1987.

[GB94] Julian W. Gardner and Philip N. Bartlett. A Brief History of Electronic Noses. *Sensors and Actuators B*, 18-19:211–220, 1994.

[GB99] Julian W. Gardner and Philip N. Bartlett. *Electronic Noses - Principles and Applications*. Oxford Science Publications, Oxford, 1999.

[GN92] M. Giurfa and J. A. Nunez. Honeybees Mark With Scent and Reject Recently Visited Flowers. *Oecologia*, 89:113–117, 1992.

[Gol98] D. Goldberg. *Genetic Algorithms in Search, Optimization and Machine Learning*. Addison Wesley, Reading, 1998.

[GT99] G. Gibson and S. J. Torr. Visual and Olfactory Responses of Haematophagous Diptera to Host Stimuli. *Medical and Veterinary Entomology*, 13:2–23, 1999.

[GW92] R. C. Gonzalez and R. E. Woods. *Digital Image Processing*. Addison-Wesley, 1992.

[Han67] W. Hangartner. Spezifität und Inaktivierung des Spurpheromons von *Lasius fuliginious Latr.* und Orientierung der Arbeiterinnen im Duftfeld. *Z. Vergl. Physiol.*, 57(2):103–136, 1967.

[Han96] Elise Hancock. A Primer on Smell. *Johns Hopkins Magazine, Electronic Edition – A Special Issue on the Senses*, 9, September 1996.

[Har54] J. D. Hartman. A Possible Objective Method for the Rapid Estimation of Flavours in Vegetables. *Proc. Am. Soc. Hort. Sci.*, 64(335), 1954.

[Hay78] P. Hayes. The Naive Physics Manifesto. In D. Michie, editor, *Expert Systems in the MicroElectronic Age*, pages 242–270. Edinburgh University Press, 1978.

[Hin75] J. O. Hinze. *Turbulence*. McGraw-Hill, New York, 1975.

[HM01] Owen Holland and David McFarland. *Artificial Ethology*. Oxford University Press, New York, 2001.

[HMG02] A.T. Hayes, A. Martinoli, and R.M. Goodman. Distributed Odor Source Localization. *IEEE Sensors Journal, Special Issue on Electronic Nose Technologies*, 2(3):260–273, 2002. June.

[HMR01] A.T. Hayes, A. Martinoli, and R.M.Goodman. Swarm Robotic Odor Localization. In *Proceedings of the IEEE/RSJ International Conference on Intelligent Robots and Systems (IROS 2001)*, volume 2, pages 1073–1078, October 2001.

[Hol75] John H. Holland. *Adaptation in Natural and Artificial Systems*. MIT Press, 1975.

[HWRG99] T. Hermle, U. Weimar, W. Rosenstiel, and W. Göpel. Performance of Selected Evaluation Methods for a Hybrid Sensor System. In *ISOEN Conference Proceedings*, pages 183–186, 1999.

[IK85] A. Ikegami and M. Kaneyasu. Olfactory Detection Using Integrated Sensors. In *Proceedings of the 3rd International Conference on Solid-State Sensors and Actuators (Transducers 1985)*, pages 136–139, New York, 1985. IEEE Press.

[IKNM96] Hiroshi Ishida, Y. Kagawa, T. Nakamoto, and Toyosaka Moriizumi. Odour-Source Localization in the Clean Room by an Autonomous Mobile Sensing System. *Sensors and Actuators B*, 33:115–121, 1996.

[IKNM99] Hiroshi Ishida, Akito Kobayashi, Takamichi Nakamoto, and Toyosaka Moriizumi. Three-Dimensional Odor Compass. *IEEE Transactions on Robotics and Automation*, 15(2):251–257, April 1999.

[IM03] Hiroshi Ishida and Toyosaka Moriizumi. Machine Olfaction for Mobile Robots. In T. C. Pearce, S. S. Schiffman, H. T. Nagle, and J. W. Gardner, editors, *Handbook of Machine Olfaction: Electronic Nose Technology*. Wiley-VCH, 2003.

[INM98] Hiroshi Ishida, T. Nakamoto, and Toyosaka Moriizumi. Remote Sensing of Gas/Odor Source Location and Concentration Distribution Using Mobile System. *Sensors and Actuators B*, 49:52–57, 1998.

[INM+01] Hiroshi Ishida, Takamichi Nakamoto, Toyosaka Moriizumi, Timo Kikas, and Jiri Janata. Plume-Tracking Robots: A New Application of Chemical Sensors. *Biol. Bull.*, 200:222–226, April 2001.

[iRo] iRobot, Burlington (USA). http://www.irobot.com (February 2004).

[ISNM94] Hiroshi Ishida, K. Suetsugu, Takamichi Nakamoto, and Toyosaka Moriizumi. Study of Autonomous Mobile Sensing System for Localization of Odor Source Using Gas Sensors and Anemometric Sensors. *Sensors and Actuators A*, 45:153–157, 1994.

[ITYM03] Hiroshi Ishida, M. Tsuruno, K. Yoshikawa, and T. Moriizumi. Spherical Gas-Sensor Array for Three-Dimensional Plume Tracking. In *Proceedings of the IEEE International Conference on Advanced Robotics*

(ICAR 2003), pages 369–374, 2003.

[IYK⁺00] Hiroshi Ishida, T. Yamanaka, N. Kushida, Takamichi Nakamoto, and Toyosaka Moriizumi. Study of Real-Time Visualization of Gas/Odor Flow Images Using Gas Sensor Array. *Sensors and Actuators B*, (65):14–16, 2000.

[Kan96] Ryohei Kanzaki. Behavioral and Neural Basis of Instinctive Behavior in Insects: Odor-Source Searching Strategies without Memory and Learning. *Robotics and Autonomous Systems*, 18:33–43, 1996.

[Kan98a] Ryohei Kanzaki. Coordination of Flipflopping Neural Signals and Head Turning During Pheromone-Mediated Walking in a Male Silkworm Moth *Bombyx mori*. *J. Comp. Physiol. A*, 183:273–282, 1998.

[Kan98b] Ryohei Kanzaki. Coordination of Wing Motion and Walking Suggests Common Control of Zigzag Motor Program in a Male Silkworm Moth. *J. Comp. Physiol. A*, 182:267–276, 1998.

[Kap01] J. Kappler. *Characterisation of High-Performance SnO_2 Gas Sensors for CO Detection by In Situ Techniques*. PhD thesis, University of Tübingen, 2001.

[KE97] Bruno Kolb and Leslie S. Ettre. *Static Headspace-Gas Chromatography - Theory and Practice*. Wiley-VCH, New York, 1997.

[Kha85] O. Khatib. Real-Time Obstacle Avoidance for Manipulators and Mobile Robots. In *Proceedings of the IEEE International Conference on Robotics and Automation (ICRA 1985)*, pages 500–505, 1985.

[KIAI87] M. Kaneyasu, A. Ikegami, H. Arima, and S. Iwanga. Smell Identification Using a Thick-Film Hybrid Gas Sensor. In *IEEE Trans. Components, Hybrids Manufact. Technol., CHMT-10*, pages 267–273, 1987.

[KL59] P. Karlson and M. Lüscher. "Pheromones" a New Term for a Class of Biologically Active Substances. *Nature*, 183:55–56, 1959.

[KLZV99] J. Kleperis, A. Lusis, J. Zubkans, and M. Veidemanis. Two Years Experience With Nordic E-Nose. In *ISOEN Conference Proceedings*, pages 11–14, 1999.

[KNSK99] Y. Kuwana, S. Nagasawa, I. Shimoyana, and R. Kanzaki. Synthesis of the Pheromone-oriented Behaviour of Silkworm Moths by a Mobile Robot with Moth Antennae as Pheromone Sensors. *Biosensors & Bioelectronics*, 14:192–202, 1999.

[KR03] Gideon Kowadlo and R. Andrew Russell. Naive Physics for Effective Odour Localisation. In *Proceedings of the Australian Conference on*

Robotics and Automation, 2003.

[Kri00]   Dietmar Krieg. *Entwicklung einer Methode zur Auswahl raumlufttechnischer Systeme mit Hilfe neuronaler Netze.* PhD thesis, Universität Stuttgart, IKE, 2000.

[KSM95]   Y. Kuwana, I. Shimoyama, and H. Miura. Steering Control of a Mobile Robot Using Insect Antennae. In *Proceedings of the IEEE/RSJ International Conference on Intelligent Robots and Systems (IROS 1995)*, pages 530–535, 1995.

[KTe]   K-Team, Switzerland. http://www.k-team.com (February 2004).

[KTN+99]   H. Kitano, S. Tadokor, H. Noda, I. Matsubara, T. Takhasi, A. Shinjou, and S. Shimada. RoboCup Rescue: Search and Rescue in Large Scale Disasters as a Domain for Multi-Agent Research. In *Proceedings of IEEE Conference on Man, Systems, and Cybernetics (SMC-99)*, 1999.

[Küh19]   A. Kühn. *Die Orientierung der Tiere im Raum.* Gustav Fischer, Jena, 1919.

[LAMdA03]   Svetlana Larionova, Nuno Almeida, Lino Marques, and A. T. de Almeida. Olfactory Coordinated Area Coverage. In *Proceedings of the IEEE International Conference on Advanced Robotics (ICAR 2003)*, pages 501–506, 2003.

[LD03a]   Achim Lilienthal and Tom Duckett. A Stereo Electronic Nose for a Mobile Inspection Robot. In *Proceedings of the IEEE International Workshop on Robotic Sensing (ROSE 2003)*, 2003.

[LD03b]   Achim Lilienthal and Tom Duckett. An Absolute Positioning System for 100 Euros. In *Proceedings of the IEEE International Workshop on Robotic Sensing (ROSE 2003)*, 2003.

[LD03c]   Achim Lilienthal and Tom Duckett. An Approach to Gas Source Localisation and Declaration by Pure Chemo-Tropotaxis. In *Autonome Mobile Systeme (AMS), 18. Fachgespräch*, pages 161–171. GDI, Informatik aktuell, 2003.

[LD03d]   Achim Lilienthal and Tom Duckett. Creating Gas Concentration Gridmaps with a Mobile Robot. In *Proc. of the 2003 IEEE/RSJ Int. Conference on Intelligent Robots and Systems (IROS 2003)*, pages 118–123, 2003.

[LD03e]   Achim Lilienthal and Tom Duckett. Experimental Analysis of Smelling Braitenberg Vehicles. In *Proceedings of the IEEE International Conference on Advanced Robotics (ICAR 2003)*, pages 375–380, 2003.

[LD03f] Achim Lilienthal and Tom Duckett. Gas Source Localisation by Constructing Concentration Gridmaps with a Mobile Robot. In *Proceedings of the European Conference on Mobile Robots (ECMR 2003)*, pages 159–164, 2003.

[LD04a] Achim Lilienthal and Tom Duckett. Building Gas Concentration Gridmaps with a Mobile Robot. *Robotics and Autonomous Systems*, 48(1):3–16, August 2004.

[LD04b] Achim Lilienthal and Tom Duckett. Experimental Analysis of Gas-Sensitive Braitenberg Vehicles. *Advanced Robotics*, 18(8):817–834, December 2004.

[Lom86] C. G. Lomas. *Fundamentals of Hot Wire Anemometry*. Cambridge University Press, 1986.

[LRZ03] Achim Lilienthal, Denis Reiman, and Andreas Zell. Gas Source Tracing With a Mobile Robot Using an Adapted Moth Strategy. In *Autonome Mobile Systeme (AMS), 18. Fachgespräch*, pages 150–160. GDI, Informatik aktuell, 2003.

[LUF+04a] Achim Lilienthal, Holger Ulmer, Holger Fröhlich, Andreas Stützle, Felix Werner, and Andreas Zell. Gas Source Declaration with a Mobile Robot. In *Proceedings of the IEEE International Conference on Robotics and Automation (ICRA 2004)*, pages 1430–1435, 2004.

[LUF+04b] Achim Lilienthal, Holger Ulmer, Holger Fröhlich, Felix Werner, and Andreas Zell. Learning to Detect Proximity to a Gas Source with a Mobile Robot. In *Proceedings of the IEEE/RSJ International Conference on Intelligent Robots and Systems (IROS 2004)*, pages 1444–1449, 2004.

[LZWW00] Achim Lilienthal, Andreas Zell, Michael R. Wandel, and Udo Weimar. Ein Autonomer Mobiler Roboter mit Elektronischer Nase. In *Autonome Mobile Systeme (AMS), 16. Fachgespräch*, pages 201–210. GDI, Informatik aktuell, 2000.

[LZWW01a] Achim Lilienthal, Andreas Zell, Michael R. Wandel, and Udo Weimar. Experiences Using Gas Sensors on an Autonomous Mobile Robot. In *Proceedings of EUROBOT 2001, 4th European Workshop on Advanced Mobile Robots*, pages 1–8. IEEE Computer Press, 2001.

[LZWW01b] Achim Lilienthal, Andreas Zell, Michael R. Wandel, and Udo Weimar. Sensing Odour Sources in Indoor Environments Without a Constant Airflow by a Mobile Robot. In *Proceedings of the IEEE International Conference on Robotics and Automation (ICRA 2001)*, pages 4005–4010, 2001.

[LZWW02] Achim Lilienthal, Andreas Zell, Michael R. Wandel, and Udo Weimar. Detektion und Lokalisierung einer Geruchsquelle mit einem autonomen mobilen Roboter. In *Robotik 2002: Leistungsstand - Anwendungen - Visionen - Trends*, volume 1679 of *VDI-Berichte*, pages 689–694, Düsseldorf, 2002. VDI/VDE-Gesellschaft Mess- und Automatisierungstechnik, VDI Verlag GmbH.

[Mar63] D. W. Marquardt. An Algorithm for Least-Squares Estimation of Nonlinear Parameters. *Journal of the Society for Industrial and Applied Mathematics*, 11:431–441, 1963.

[May79] P. S. Maybeck. *Stochastic Models, Estimation, and Control, Volume 1*, volume 1. Academic Press, 1979.

[MCH+00] R. Murphy, J. Casper, J. Hyams, M. Micire, and B. Minten. Mobility and Sensing Demands in USAR. In *Proceedings of IECON 2000 (invited)*, 2000.

[McN87] Michael J. McNish. Effects of uniform target density on random search. Master's thesis, Naval Postgraduate School, Monterey, California, 1987.

[MEC92] J. Murlis, J. S. Elkington, and R. T. Carde. Odor Plumes and How Insects Use Them. *Annual Review of Entomology*, 37:505–532, 1992.

[MK99] Graham A. Mann and George Katz. Chemical Trail Guidance for Floor Cleaning Machines. In *Proceedings of the 2nd International Conference on Field & Service Robotics (FSR 1999)*, pages 195–204, August 1999.

[MM96] Martin C. Martin and Hans P. Moravec. Robot Evidence Grids. Technical Report CMU-RI-TR-96-06, The Robotics Institute, Carnegie Mellon University, 1996.

[Mon61] R. W. Moncrieff. An Instrument for Measuring and Classifying Odours. *J. Appl. Physiol.*, 16:742–749, 1961.

[Mor95] Toyosaka Moriizumi. Biomimetic Sensing Systems With Arrayed Nonspecific Sensors. In *Proceedings of the 8th International Conference on Solid-State Sensors and Actuators (Transducers 1995)*, pages 39–42, 1995.

[NBF96] Bradford Nichols, Dick Buttlar, and Jackie Farrell. *Pthreads Programming*. O'Reilly, 1996.

[NIM96] Takamichi Nakamoto, Hiroshi Ishida, and Toyosaka Moriizumi. An Odor Compass for Localizing an Odor Source. *Sensors and Actuators B*, 35:32–36, 1996.

[NIM99]   Takamichi Nakamoto, Hiroshi Ishida, and Toyosaka Moriizumi. A
           Sensing System for Odor Plumes. *Analytical Chem. News & Features*,
           1:531–537, August 1999.

[OMO⁺92]   R. Olafsson, E. Martinsdottir, G. Olafsdottir, S. I. Sigfusson, and J. W.
           Gardner. Monitoring of Fish Freshness Using Tin Oxide Sensors. In
           Sensors and Sensory Systems for an Electronic Nose, pages 257–272,
           1992.

[Pap]   ebm-papst, Germany. http://www.papst.de (February 2004).

[PCLL⁺01] M. C. Pinto, D. H. Campbell-Lendrum, A. L. Lozovei, U. Teodoro,
           and C. R. Davies. Phlebotomine Sandfly Responses to Carbon Dioxide
           and Human Odour in the Field. *Medical and Veterinary Entomology*,
           15:132–139, 2001.

[PD82]   K. Persaud and G. Dodd. Analysis of Discrimination Mechanisms
           of the Mammalian Olfactory System Using a Model Nose. *Nature*,
           (299):352–355, 1982.

[Pen99]   David G. Penny. *Environmental Health Criteria, No. 213: Carbon
           Monoxide*. World Health Organisation (WHO), 2. edition, 1999.

[PFTV92]   William H. Press, Brian P. Flannery, Saul A. Teukolsky, and
           William T. Vetterling. *Numerical Recipes: The Art of Scientific Com-
           puting*. Cambridge University Press, Cambridge (UK) and New York,
           2nd edition, 1992.

[PH00]   E. Prigge and J. P. How. An Indoor Absolute Positioning System
           with No Line of Sight Restrictions and Building-Wide Coverage. In
           *Proceedings of the IEEE International Conference on Robotics and Au-
           tomation (ICRA 2000)*, pages 1015–1022, 2000.

[PNDW98]   D. Pagac, M. E. Nebot, and H. Durrant-Whyte. An Evidential Ap-
           proach to Map Building for Autonomous Vehicles. *IEEE Transactions
           on Robotics and Automation*, 14(4):623–629, 1998.

[Pol]   Vermont Polhemus, Colchester. Liberty product brochure.
           http://www.polhemus.com/LIBERTY/lib_brochure.pdf (Feb. 2004).

[PSM86]   M. Pluijm, G. Sars, and C. H. Massen. Calibration Unit for Micro-
           Anemometers at Very Low Air Velocities. *Appl. Sci. Res.*, 43:227–234,
           1986.

[PWV86]   E. Priesner, P. Witzgall, and S. J. Voerman. Field Attraction Response
           of Raspberry Clearwing Moths, *Pennisethia hylaeiformis* Lasp. (Lep-
           idoptera: Sesiidae), to Candidate Pheromone Chemicals. *Journal of
           Applied Entomology*, 102:195–210, 1986.

[RBHSW03] R. Andrew Russell, Alireza Bab-Hadiashar, Rod L. Shepherd, and Gordon G. Wallace. A Comparison of Reactive Chemotaxis Algorithms. *Robotics and Autonomous Systems*, 45:83–97, 2003.

[RJB97] J. Rumbaugh, I. Jacobson, and G. Booch. *Unified Modeling Language Reference Manual*. Addison-Wesley Longman, 1997.

[RK00] R. A. Russell and S. Kennedy. A Novel Airflow Sensor for Miniature Mobile Robots. *Mechatronics*, 10(8):935–942, 2000.

[RK03] S. Rajasekharan and C. Kambhampati. The Current Opinion on the Use of Robots for Landmine Detection. In *Proceedings of the IEEE International Conference on Robotics and Automation (ICRA 2003)*, pages 4252–4257, 2003.

[RKK00] R. A. Russell, L. Kleeman, and S. Kennedy. Using Volatile Chemicals to Help Locate Targets in Complex Environments. In *Proceedings of the Australian Conference on Robotics and Automation*, pages 87–91, Melbourne, 2000.

[RM86] D. E. Rumelhart and J. L. McClelland. *Parallel Distributed Processing: Explorations in the Microstructure of Cognition, Volume 1: Foundations*. MIT Press, 1986.

[RM01] Cliff Randell and Henk L. Muller. Low Cost Indoor Positioning System. In *International Conference on Ubiquitous Computing UbiComp 2001*, pages 42–48, 2001.

[RMV91] Roberto Rozas, Jorge Morales, and Daniel Vega. Artificial Smell Detection for Robotic Navigation. In *Proceedings of the IEEE International Conference on Robotics and Automation (ICRA 1991)*, pages 1730–1733, 1991.

[RP02] R. Andrew Russell and Anies H. Purnamadjaja. Odour and Airflow: Complementary Senses for a Humanoid Robot. In *Proceedings of the IEEE International Conference on Robotics and Automation (ICRA 2002)*, pages 1842–1847, 2002.

[RTDMS95] R. Andrew Russell, David Thiel, Reimundo Deveza, and Alan Mackay-Sim. A Robotic System to Locate Hazardous Chemical Leaks. In *Proceedings of the IEEE/RSJ International Conference on Intelligent Robots and Systems (IROS 1995)*, pages 556–561, 1995.

[RTMS94] R. Andrew Russell, David Thiel, and Alan Mackay-Sim. Sensing Odour Trails for Mobile Robot Navigation. In *Proceedings of the IEEE International Conference on Robotics and Automation (ICRA 1994)*, pages 2672–2677, 1994.

[Rus95]   R. Andrew Russell. Laying and Sensing Odor Markings as a Strat-
          egy for Assisting Mobile Robot Navigation Tasks. *IEEE Robotics &
          Automation Magazine*, pages 3–9, September 1995.

[Rus99a] R. Andrew Russell. Ant Trails - an Example for Robots to Follow?
          In *Proceedings of the IEEE International Conference on Robotics and
          Automation (ICRA 1999)*, pages 2698–2703, 1999.

[Rus99b] R. Andrew Russell. *Odour Sensing for Mobile Robots*. World Scientific,
          1999.

[RW02]   P. J. W. Roberts and D. R. Webster. Turbulent Diffusion. In H. Shen,
          A. Cheng, K.-H. Wang, M.H. Teng, and C. Liu, editors, *Environmen-
          tal Fluid Mechanics - Theories and Application*. ASCE Press, Reston,
          Virginia, 2002.

[Sau59]   G. Sauerbrey. Verwendung von Schwingquarzen zur Wägung dünner
          Schichten und zur Mikrowägung. *Zeitschrift für Physik*, 155:206–222,
          1959.

[SBJW02] Alessandro Safiotti, A. Björklund, S. Johansson, and Zbigniew Wasik.
          Team Sweden (Team Description). *RoboCup 2001: Robot Soccer World
          Cup V*, pages 725–729, 2002.

[Sch69]   D. Schneider. Insect Olfaction: Deciphering System for Chemical Mes-
          sages. *Science*, 163:1031–1037, 1969.

[SIC]   SICK, Düsseldorf (Germany). http://www.sick.de (February 2004).

[SLV93]   G. Sandini, G. Lucarini, and M. Varoli. Gradient-Driven Self-
          Organizing Systems. In *Proceedings of the IEEE/RSJ International
          Conference on Intelligent Robots and Systems (IROS 1993)*, pages 429–
          432, July 1993.

[SM01]   W. D. Smyth and J. N. Moum. 3D Turbulence. In John Steele, Steve
          Thorpe, and Karl Turekian, editors, *Encyclopedia of Ocean Sciences*.
          Academic Press, 2001.

[Smi78]   A. R. Smith. Color Gamut Transform Pairs. *ACM SIGGRAPH Com-
          puter Graphics*, 12(3):12–19, 1978.

[SMVD95] E. Stella, F. Musio, L. Vasanelli, and A. Distante. Goal-oriented Mo-
          bile Robot Navigation Using an Odour Sensor. In *Proceedings of the
          Intelligent Vehicles Symposium 1995*, pages 147–151, 1995.

[Som01]   Ian Sommerville. *Software Engineering*. Addison-Wesley, 6th edition,
          2001.

[SON]   SONY, Japan. http://www.sony.com (February 2004).

[SSRB00] Douglas Schmidt, Michael Stal, Hans Rohnert, and Frank Buschmann. *Pattern-Oriented Software Architecture - Patterns for Concurrent and Networked Objects*, volume 2. Wiley, Sussex, England, 2000.

[Sud67] J. H. Sudd. *An Introduction to the Behaviour of Ants*. Arnold Pub. Ltd, London, 1967.

[SW98] T. Sharpe and B. Webb. Simulated and Situated Models of Chemical Trail Following in Ants. In R. Pfeifer, B. Blumberg, J.-A. Meyer, and S.W. Wilson, editors, *Proceedings of the 5th Conference on Simulation of Adaptive Behaviour*, pages 195–204, Zürich, Switzerland, 1998.

[Vap95] V. Vapnik. *The Nature of Statistical Learning Theory*. Springer, New York, NY, 1995.

[Vid] Videre Design, USA. http://www.videredesign.com (February 2004).

[Vog97] H. Vogel. *Gerthsen Physik*. Springer Verlag, Berlin, 19. edition, 1997.

[WH64] W. F. Wilkens and J. D. Hartman. An Electronic Analogue for the Olfactory Process. *Ann. NY Acad. Sci.*, 116:608–612, 1964.

[WHSL93] F. Winquist, E.G. Hornsten, H. Sundgren, and I. Lundström. Performance of an Electronic Nose for Quality Estimation of Ground Meat. *Measurement Science and Technology*, 4:1493–1500, 1993.

[Wil] Mark A. Willis. Modeling Approaches to Understand Odor-Guided Locomotion. http://flightpath.neurobio.arizona.edu/Model (February 2004).

[WK01] Robert A. Wilson and Frank C. Keil, editors. *The MIT Encyclopedia of the Cognitive Sciences*. Bradford Books, 2001.

[WLD+03] Michael R. Wandel, Achim Lilienthal, Tom Duckett, Udo Weimar, and Andreas Zell. Gas Distribution in Unventilated Indoor Environments Inspected by a Mobile Robot. In *Proc. of the IEEE Int. Conference on Advanced Robotics (ICAR 2003)*, pages 507–512, 2003.

[WLZW02] Michael Wandel, Achim Lilienthal, Andreas Zell, and Udo Weimar. Mobile Robot Using Different Senses. In *ISOEN Conference Proceedings*, pages 128–130, 2002.

[WWLZ01] Michael R. Wandel, Udo Weimar, Achim J. Lilienthal, and Andreas Zell. Leakage Localisation with a Mobile Robot Carrying Chemical Sensors. In *International IEEE Conference on Electronics, Circuits, and Systems (ICECS)*, pages 1247–1250. GDI, 2001.